REAL OPTIONS IN ENGINEERING DESIGN, OPERATIONS, AND MANAGEMENT

REAL OPTIONS IN
ENGINEERING DESIGN, OPERATIONS, AND MANAGEMENT

Edited by
Harriet Black Nembhard
Mehmet Aktan

CRC Press
Taylor & Francis Group
Boca Raton London New York

CRC Press is an imprint of the
Taylor & Francis Group, an **informa** business

CRC Press
Taylor & Francis Group
6000 Broken Sound Parkway NW, Suite 300
Boca Raton, FL 33487-2742

International Standard Book Number: 978-1-4200-7169-6 (Hardback)

Library of Congress Cataloging-in-Publication Data

Real options in engineering design, operations, and management / Harriet Black
    Nembhard, Mehmet Aktan.
        p. cm.
    Includes bibliographical references and index.
    ISBN 978-1-4200-7169-6 (hardcover : alk. paper)
    1. Engineering design--Management. 2. Operations research. 3. Real options
(Finance) I. Nembhard, Harriet Black. II. Aktan, Mehmet. III. Title.

    TA174.R376 2010
    620'.0042--dc22   2009031430

**Visit the Taylor & Francis Web site at
http://www.taylorandfrancis.com**

**and the CRC Press Web site at
http://www.crcpress.com**

Contents

Preface

Chapters 1 through 3 provide an introduction to real options and an overview of how real options came to be used as an analytical tool in engineering problems. Chapters 4 through 6 deliver a real options perspective on operations. Chapters 7 through 10 cover a real options perspective on design. Chapters 11 through 15 discuss a real options perspective on management. In particular, the last chapter shows how easy it is to use real options software for the business problems that may arise out of engineering considerations.

Many people have contributed to the production of this book. In addition to each chapter author, the editorial team at CRC Press including Cindy Carelli and Catherine Giacari has been enormously supportive.

We welcome comments on the book from readers.

Harriet Black Nembhard
Pennsylvania State University

Mehmet Aktan
Atatürk University

Editors

Harriet Black Nembhard, PhD, is an associate professor of industrial engineering at Pennsylvania State University. She is also the director of the Laboratory for Quality Engineering and System Transitions, which examines ways to improve decision making in operations for periods and points of change. Dr. Nembhard is the author or co-author of more than 45 research publications, has received research support from the National Science Foundation, and sits on the editorial boards of four research journals. Dr. Nembhard earned her PhD from the University of Michigan. She is an elected member of the International Academy for Quality (IAQ) and a member of the Institute of Industrial Engineers (IIE), and the Institute for Operations Research and Management Science (INFORMS).

Mehmet Aktan, PhD, is an assistant professor of industrial engineering at Atatürk University, Turkey. He received his PhD in industrial engineering from the University of Wisconsin–Madison, his MS from the University of Iowa, and his BS from Bogaziçi University. He is the author and co-author of several research publications in the areas of financial engineering, statistical methods, and human factors.

Contributors

Mehmet Aktan
Department of Industrial Engineering
Faculty of Engineering
Atatürk University
Erzurum, Turkey
maktan@atauni.edu.tr

Anteneh Ayanso
Department of Finance, Operations
    and Information Systems
Faculty of Business
Brock University
St. Catharines, Ontario, Canada

David N. Ford
Zachry Department of Civil
    Engineering
Texas A&M University
College Station, Texas
dford@civil.tamu.edu

Michael Garvin
Myers-Lawson School of Construction
Virginia Technical Institute and State
    University
Blacksburg, Virginia
garvin@vt.edu

Hemantha Herath
Department of Accounting
Faculty of Business
Brock University
St. Catharines, Ontario, Canada
HHerath@brocku.ca

Konstantinos Kalligeros
London, United Kingdom
kkalligeros@gmail.com

Vassilios N. Kazakidis
School of Engineering
Laurentian University
Sudbury, Ontario, Canada
vkazakidis@laurentian.ca

Min-Jung Kim
Harold and Inge Marcus Department
    of Industrial and Manufacturing
    Engineering
Pennsylvania State University
University Park, Pennsylvania
mzk148@psu.edu

Zachary Mayer
School of Engineering
Laurentian University
Sudbury, Ontario, Quebec

Johnathan Mun
Real Options Valuation, Inc.
Dublin, California
jcmun@realoptionsvaluation.com

David Nembhard
Harold and Inge Marcus Department
    of Industrial and Manufacturing
    Engineering
Pennsylvania State University
University Park, Pennsylvania
DNembhard@psu.edu

Harriet Black Nembhard
Harold and Inge Marcus Department
    of Industrial and Manufacturing
    Engineering
Pennsylvania State University
University Park, Pennsylvania
HBNembhard@psu.edu

Richard de Neufville
Technology and Policy Program
Massachusetts Institute of
    Technology
Cambridge, Massachusetts
Ardent@mit.edu

Ruwen Qin
Capital One Financial Service Inc.
McLean, Virginia
ruwen.qin@capitalone.com

Stefan Scholtes
Judge Business School
University of Cambridge
Cambridge, United Kingdom

Leyuan Shi
Department of Industrial and Systems
    Engineering
University of Wisconsin–Madison
Madison, Wisconsin
leyuan@engr.wisc.edu

Olivier de Weck
Engineering Systems Division
Massachusetts Institute of Technology
Cambridge, Massachusetts

Julie Ann Stuart Williams
Department of Management and MIS
University of West Florida
Pensacola, Florida
JAWilliams@uwf.edu

1 Introduction

Harriet Black Nembhard
Pennsylvania State University

Mehmet Aktan
Atatürk University

CONTENTS

In today's fast-paced business environment, flexibility provides potentially great strategic benefits to firms. Flexibility allows firms to compete more effectively in a world of substantial price and demand uncertainty, product variety, short product life cycles, and rapid product development.

The term "real options" was coined by Stewart Myers in 1977. He argued that the value of a firm includes the real assets in place plus the present value of options to make further investments in the future. These future investment opportunities are undertaken at the discretion of the firm, just as options on trade are exercised only when it is profitable to do so.

In other words, the real options approach seeks to quantify just how valuable flexibility really is. It takes into account the fact that most investments have managerial decisions and manufacturing uncertainty embedded within them. This method attempts to incorporate both the uncertainty inherent in business and the active decision making required for a strategy to succeed.

Fundamentally, an option is the right, but not the obligation, to take an action in the future. With real options the initial investment related to an asset buys the potential opportunity to continue, expand, or abandon the use of the asset when it is favorable to do so, but does not carry the obligation to realize some losses when unfavorable conditions prevail. Managerial decisions lead to strategic and financial goals and they can follow different paths. They usually come in incremental steps. The real option at each step in the decision-making process is the freedom of choice to embark on the next step in the climb, or to choose against doing so based on the examination of additional information. An investment decision is rarely a now-or-never decision and rarely a decision that cannot be abandoned or changed. In most instances, the decision can be delayed or accelerated, and often it comes in sequential steps with various decision points, including "go" and "no-go" alternatives. All of these choices are real managerial options and impact on the value of the investment opportunity. Further, managers are very conscious of preserving a certain freedom of choice to respond to future uncertainties (Brach 2003). When executives create strategy, they project themselves and their organizations into the future, creating a path from where they are now to where they want to be some years down the road. But a plan cannot be formulated and followed mindlessly. It is intuitively understood that flexibility has value.

Real options may be categorized into three broad categories, as Copeland and Keenan (1998a, 1998b) described:

1. *Investment/growth options.* These include (1) scale-up options, where early entrants can scale up later through sequential investments as their market grows; (2) switch-up options, where speedy commitment to the first generation of a product or technology gives a firm preferential position to switch to the next generation of the product or technology; and (3) scope-up options, where investments in proprietary assets in one industry enable a firm to enter another industry cost-effectively.
2. *Deferral/learning options.* Also called study/start options, these are opportunities to delay investment until more information or skill is acquired. For example, a pharmaceutical firm uses real options analysis (ROA) to evaluate drug development projects, in which investments are made in several phases of experimentation with the drug compound before seeking regulatory approval and going to market.
3. *Disinvestment/shrinkage options.* These include (1) scale-down options, where new information that changes the expected payoffs can cause managers to shrink or shut down a project before completion; (2) switch-down options, where managers have the ability to switch to more cost-effective and flexible assets as new information is obtained; and (3) scope-down options, where operations are decreased or even abandoned when there is no further potential in a business opportunity.

The next issue that is critical to managers and financial planners is to understand and capture *precisely* the value of this flexibility. Achieving this with traditional financial tools, such as net present value (NPV) analysis and discounted cash flow (DCF) methods, is difficult because the value of switching depends on the current status, and once you have switched, the value of switching depends on your new status and the switching options that it has. NPV and DCF cannot keep track of those interdependencies. In financial terms, a business strategy is much more like a series of options than like a series of static cash flows. Options analysis can deliver extra insight. Simple option pricing for exchange-traded puts and calls is fairly straightforward, and many authors present the basics lucidly (e.g., Hull 2008; Luenberger 1998). Real options build upon these ideas to quantify opportunities that are not financial instruments.

An excellent and practical reference is *Real Options Analysis: Tools and Techniques,* 2nd ed. (Wiley Finance; Mun 2006). Chapter 2, written by the author of that book, gives the reader an overview of many of the issues, criticisms, and viewpoints surrounding the use of real options in practice.

From an engineering perspective, manufacturing flexibility has been one of the key drivers of further work in real options. The origins of this work are discussed in Chapter 3. For other references on real options from a more theoretical point of view, we recommend Amram and Kulatilaka (1999), Trigeorgis (1999), and Copeland and Antikarov (2001).

The goal of this book is to present the state of the art in real options for engineering design, operations, and management. Therefore, Chapters 4–14 of this volume investigate a wide spectrum of fieldwork as follows.

MANUFACTURING OPERATIONS

A manufacturing firm may delay, expand, contract, switch, or abandon a manufacturing project. Chapter 4 uses several scenarios that are valued over multiple periods.

QUALITY IMPROVEMENT

Chapter 5 presents a real options framework to value a quality improvement program and to determine strategic decisions about the program. A multinomial lattice technique is used to value the expected benefit potential of the quality improvement program.

PRODUCT OUTSOURCING

Chapter 6 investigates the value of outsourcing flexibility considering volatile currency exchange rates between client and vendor countries. The strategy is selected among supplying and/or producing at the home country, and outsourcing offshore.

ARCHITECTURE/ENGINEERING/CONSTRUCTION

The limited adoption and use of real options by practicing managers in the architecture/engineering/construction (AEC) industry remain an important challenge. Chapter 7 describes a risk-rich managerial practice in which real options can add value but are not fully exploited. This setting is used as a basis for identifying and describing specific barriers to widespread real options adoption and use by practicing project managers. These barriers are used to suggest tools, methods, and approaches that may reduce those barriers.

IMPROVING THE DESIGN OF ENGINEERING SYSTEMS

Flexible designs can deliver benefits because their underlying architecture enables managers to adapt projects to circumstances that develop. Owners can thus cut losses by avoiding undesirable outcomes, and increase gains by taking advantage of new opportunities. Chapter 8 focuses on the development of valuable flexibility in designs.

MINING SYSTEMS PLANNING AND DESIGN

Large multifaceted capital projects in the mineral resource industry are often associated with diverse sources of performance uncertainty. The introduction of flexibility into the planning process is required to counter the downturns and provide the ability to exploit the upturns that can develop over the life of a mining production system. The methods to assess the risk associated with a particular mining process and the flexible alternatives considered are discussed in Chapter 9, and a methodology is applied to case studies from underground mines in Canada.

ENGINEERING SYSTEMS DESIGN

Chapter 10 examines how designers can create real options in flexible systems, and how they can compare or even optimize the flexibility among alternative design solutions. A necessary element in this process is the consistent comparison of the risk exposure achieved by alternative designs, flexible or not.

WORKFORCE CROSS-TRAINING

Workforce cross-training involves a dynamic investment on workforce flexibility. Chapter 11 proposes a real options framework that models the cross-training policy as an approximation of an American call option using binomial lattices. Value stems from the merit of dynamic cross-training compared with the deterministic case using traditional discounted cash flow techniques. The work is discussed in the context of a volatile production system characterized by product dynamics, labor dynamics, task heterogeneity, and workforce heterogeneity.

SUSTAINABLE PRODUCT QUALITY MANAGEMENT

Chapter 12 considers a company that has the flexibility of producing both ordinary and more sustainable "green" products, and is striving to improve its overall quality. It presents a model to evaluate the optimal strategies that will maximize the expected profit using real options analysis. It illustrates the use and sensitivity of the model for desktop computer production, and discusses future sustainable product quality attributes and their product life cycle management implications for the electronics industry.

NANOTECHNOLOGY

Nanotechnology increasingly shows potential in research and development, as well as some commercial applications. The U.S. government has spent over $1 billion in promoting nanotech initiatives. Chapter 13 explores the opportunities and challenges associated with making good decisions for nanotechnology investments from the perspective of real options.

PHARMACEUTICAL DEVELOPMENT

New drug development has uncertainty at every project phase. Chapter 14 explores a method for deciding upon the optimal timing for partnerships between labs and investors. In the proposed model, the decision to invest in a new drug development as an aspect of a pharmaceutical company's consideration is represented as exercising a call option. A decision to sell ownership of a new drug is considered as a biotechnology company exercising a put option. The optimal timing depends upon a number of specific investment factors.

After these forward-reaching applications are presented, we conclude with chapter 15, which demonstrates how to model real options using the Super Lattice Solver (SLS) software tool.

We hope that this book expands the understanding of real options in the engineering domain and inspires a broader consideration of research impact through the real options decision-making lens.

REFERENCES

Amram, M., and Kulatilaka, N. 1999. *Real options: Managing strategic investment in an uncertain world.* Boston: Harvard Business School Press.

Brach, M. A. 2003. *Real options in practice.* Hoboken, NJ: John Wiley.

Copeland, T., and Antikarov, V. 2001. *Real options: A practitioner's guide.* New York: Texere.

Copeland, T. E., and Keenan, P. T. 1998a. How much is flexibility worth? *The McKinsey Quarterly,* (2): 38–49.

Copeland, T. E., and Keenan, P. T. 1998b. Making real options real. *The McKinsey Quarterly,* (3): 129–141.

Hull, J. C. 2008. *Options, futures, and other derivatives,* 7th ed. Upper Saddle River, NJ: Prentice Hall.

Luenberger, D. G. 1998. *Investment Science.* New York: Oxford University Press.

Mun, J. 2006. *Real options analysis: Tools and techniques,* 2nd ed. New York: Wiley Finance.

Myers, S. C. 1977. Determinants of corporate borrowing. *Journal of Financial Economics* 5:147–175.

Trigeorgis, L. 1999. *Real options: Managerial flexibility and strategy in resource allocation.* Cambridge, MA: MIT Press.

2 Real Options in Practice

Johnathan Mun
Real Options Valuation, Inc.

CONTENTS

Business conditions are fraught with uncertainty and risks. These uncertainties hold with them valuable information. When uncertainty becomes resolved through the passage of time, managers can make the appropriate midcourse corrections through a change in business decisions and strategies. This chapter gives an overview on how real options incorporate this learning model, akin to having a strategic road map, whereas traditional analyses that neglect this managerial flexibility will grossly undervalue certain projects and strategies. It also introduces the concepts in implementing real options, as well as a balanced perspective on what is really needed to do so.

2.1 WHAT ARE REAL OPTIONS?

In the past, corporate investment decisions were cut-and-dried. Buy a new machine that is more efficient, make more products costing a certain amount, and if the benefits outweigh the costs, execute the investment. Hire a larger pool of sales associates, expand the current geographical area, and if the marginal increase in forecast sales revenues exceeds the additional salary and implementation costs, start hiring. Need a new manufacturing plant? Show that the construction costs can be recouped quickly and easily by the increase in revenues the plant will generate through new and improved products, and the initiative is approved.

However, real-life business conditions are a lot more complicated. Your firm decides to go with an e-commerce strategy, but multiple strategic paths exist. Which path do you choose? What are the options you have? If you choose the wrong path, how do you get back on the right track? How do you value and prioritize the paths that exist? You are a venture capitalist firm with multiple business plans to consider. How do you value a start-up firm with no proven track record? How do you structure a mutually beneficial investment deal? What is the optimal timing to a second or third round of financing? Real options are useful in valuing a firm through its strategic business options.

Real options are also useful as a strategic business tool in capital investment decisions. For instance, should a firm invest millions in a new e-commerce initiative? How does a firm choose among several seemingly cashless, costly, and unprofitable information technology infrastructure projects? Should a firm invest its billions in a risky research and development initiative? The consequences of a wrong decision can be disastrous or even terminal for certain firms. In a traditional discounted cash flow model, these questions cannot be answered with any certainty. In fact, some of the answers generated through the use of the traditional discounted cash flow model are flawed because the model assumes a static, one-time decision-making process, whereas the real options approach takes into consideration the strategic managerial options certain projects create under uncertainty and management's flexibility in exercising or abandoning these options at different points in time, when the level of uncertainty has decreased or has become known over time.

The real options approach incorporates a learning model, such that management makes better and more informed strategic decisions when some levels of uncertainty are resolved through the passage of time. The discounted cash flow analysis assumes a static investment decision and assumes that strategic decisions are made initially with no recourse to choose other pathways or options in the future. To create a good analogy of real options, visualize it as a strategic road map of long and winding roads with multiple perilous turns and branches along the way. Imagine the intrinsic and extrinsic value of having such a road map or global positioning system when navigating through unfamiliar territory, as well as having road signs at every turn to guide you in making the best and most informed driving decisions. Such a strategic map is the essence of real options.

The answer to evaluating such projects lies in real options analysis, which can be used in a variety of settings, including pharmaceutical drug development, oil and gas exploration and production, manufacturing, start-up valuation, venture capital investment, information technology infrastructure, research and development, mergers and acquisitions, e-commerce and e-business, intellectual capital development, technology development, facility expansion, business project prioritization, enterprise-wide risk management, business unit capital budgeting, licenses, contracts, intangible asset valuation, and the like. The following section illustrates some business cases and how real options can assist in identifying and capturing additional strategic value for a firm.

2.2   THE REAL OPTIONS SOLUTION IN A NUTSHELL

Simply defined, real options comprise a systematic approach and integrated solution using financial theory, economic analysis, management science, decision sciences, statistics, and econometric modeling in applying options theory in valuing real physical assets, as opposed to financial assets, in a dynamic and uncertain business environment where business decisions are flexible in the context of strategic capital investment decision making, valuing investment opportunities, and project capital expenditures.

Real options are crucial in the following situations:

- Identifying different corporate investment decision pathways or projects that management can navigate given highly uncertain business conditions
- Valuing each of the strategic decision pathways and what it represents in terms of financial viability and feasibility
- Prioritizing these pathways or projects based on a series of qualitative and quantitative metrics
- Optimizing the value of strategic investment decisions by evaluating different decision paths under certain conditions or using a different sequence of pathways that can lead to the optimal strategy
- Timing the effective execution of investments and finding the optimal trigger values and cost or revenue drivers
- Managing existing or developing new optionalities and strategic decision pathways for future opportunities

Strategic options do have significant intrinsic value, but this value is realized only when management decides to execute the strategies. Real options theory assumes that management is logical and competent and that management acts in the best interests of the company and its shareholders through the maximization of wealth and minimization of risk of losses. For example, suppose a firm owns the rights to a piece of land that fluctuates dramatically in price. An analyst calculates the volatility of prices and recommends that management retain ownership for a specified time period, where within this period there is a good chance that the price of real estate will triple. Therefore, management owns a call option, an *option to wait* and defer sale for a particular time period. The value of the real estate is therefore higher than the value that is based on today's sale price. The difference is simply this option to wait. However, the value of the real estate will not command the higher value if prices do triple but management decides not to execute the option to sell. In that case, the price of real estate goes back to its original levels after the specified period, and then management finally relinquishes its rights. *Strategic optionality value can be obtained only if the option is executed; otherwise, all the options in the world are worthless.*

Was the analyst right or wrong? What was the true value of the piece of land? Should it have been valued at its explicit value on a deterministic case where you know what the price of land is right now, and therefore this is its value; or should it include some types of optionality where there is a good probability that the price of land could triple in value and, hence, the piece of land is truly worth more than it is now and should therefore be valued accordingly? The latter is the real options view. The additional strategic optionality value can be obtained only if the option is executed; otherwise, all the options in the world are worthless. This idea of *explicit* versus *implicit* value becomes highly significant when management's compensation is tied directly to the actual performance of particular projects or strategies.

To further illustrate this point, suppose the price of the land in the market is currently $10 million. Further, suppose that the market is highly liquid and volatile and that the firm can easily sell off the land at a moment's notice within the next five years, the same amount of time the firm owns the rights to the land. If there is a 50 percent chance the price will increase to $15 million and a 50 percent chance it will decrease to $5 million within this time period, is the property worth an expected value of $10 million? If the price rises to $15 million, management should be competent and rational enough to execute the option and sell that piece of land immediately to capture the additional $5 million premium. However, if management acts inappropriately or decides to hold off selling in the hopes that prices will rise even further, the property value may eventually drop back down to $5 million. Now, how much is this property really worth? What if there happens to be an *abandonment option?* Suppose there is a perfect counterparty to this transaction who decides to enter into a contractual agreement whereby, for a contractual fee, the counterparty agrees to purchase the property for $10 million within the next five years, regardless of the market price and executable at the whim of the firm that owns the property. Effectively, a safety net has been created whereby the minimum floor value of the property has been set at $10 million (less the fee paid). That is, there is a limited downside but an unlimited upside, as the firm can always sell the property at market price if it exceeds the floor value. Hence, this strategic *abandonment option* has increased the value of the property significantly. Logically, with this *abandonment option* in place, the value of the land with the option is definitely worth more than $10 million. The real options approach seeks to value this additional inherent flexibility. Real options analysis allows the firm to determine how much this safety downside insurance or abandonment option is worth (i.e., what is the fair market value of the contractual fee to obtain the option?), the optimal trigger price (i.e., what price will make it optimal to sell the land?), and the optimal timing (i.e., what is the optimal amount of time to hold on to the land?).

2.3   IMPLEMENTING REAL OPTIONS ANALYSIS

First, it is vital to understand that real options analysis is *not* a simple set of equations or models. It is an *entire decision-making process* that enhances the traditional decision analysis approaches. It takes what has been tried-and-true financial analytics and evolves it to the next step by pushing the envelope of analytical techniques. In addition, it is vital to understand that 50 percent of the value in real options analysis is simply thinking about it. Another 25 percent of the value comes from the number-crunching activities, while the final 25 percent comes from the results interpretation and explanation to management. Several issues should be considered when attempting to implement real options analysis:

- *Tools:* The correct tools are important. These tools must be more comprehensive than initially required because analysts will grow into them over time. Do not be restrictive in choosing the relevant tools. Always provide room for expansion. Advanced tools will relieve the analyst of detailed model building and let him or her focus instead on 75 percent of the value—thinking about the problem and interpreting the results. Chapter 16 of this book further discusses the Real Options Super Lattice Solver (SLS) software and how even complex and customized real options problems can be solved with great ease.

- *Resources:* The best tools in the world are useless without the relevant human resources to back them up. Tools do not eliminate the analyst, but enhance the analyst's ability to effectively and efficiently execute the analysis. The right people with the right tools will go a long way. Because there are only a few true real options experts in the world who truly understand the theoretical underpinnings of the models as well the practical applications, care should be taken in choosing the correct team. A team of real options experts is vital in the success of the initiative. A company should consider building a team of in-house experts to implement real options analysis and to maintain the ability for continuity, training, and knowledge transfer over time. Knowledge and experience in the theories, implementation, training, and consulting are the core requirements of this team of individuals.
- *Senior management buy-in:* The analysis buy-in has to be top-down where senior management drives the real options analysis initiative. A bottom-up approach where a few inexperienced junior analysts try to impress the powers that be will fail miserably, or it will be passed off as another business school fad and not applicable to "real life."

2.4 CRITICISMS, CAVEATS, AND MISUNDERSTANDINGS IN REAL OPTIONS

Before embarking on a real options analysis, analysts should be aware of several caveats. The following five requirements need to be satisfied before a real options analysis can be run:

- *A financial model must exist.* Real options analysis requires the use of an existing discounted cash flow model, as real options build on the existing tried-and-true approaches of current financial-modeling techniques. If a model does not exist, it means that strategic decisions have already been made and no financial justifications are required, and hence, there is no need for financial modeling or real options analysis.
- *Uncertainties must exist.* Otherwise, the option value is worthless. If everything is known for certain in advance, then a discounted cash flow model is sufficient. In fact, when volatility (a measure of risk and uncertainty) is zero, everything is certain, the real options value is zero, and the total strategic value of the project or asset reverts to the net present value in a discounted cash flow model.
- *Uncertainties must affect decisions when the firm is actively managing the project, and these uncertainties must affect the results of the financial model.* These uncertainties will then become risks, and real options can be used to hedge the downside risk and take advantage of the upside uncertainties.
- *Management must have strategic flexibility or options to make midcourse corrections when actively managing the projects.* Otherwise, do not apply real options analysis when there are no options or management flexibility to value.
- *Management must be smart enough and credible enough to execute the options when it becomes optimal to do so.* Otherwise, all the options in the

world are useless unless they are executed appropriately, at the right time, and under the right conditions.

There are also several criticisms against real options analysis. It is vital that the analyst understands what they are and what the appropriate responses are, prior to applying real options.

- *Real options analysis is merely an academic exercise and is not practical in actual business applications.* Nothing is further from the truth. Although it was true in the past that real options analysis was merely academic, many corporations have begun to embrace and apply real options analysis. Also, its concepts are very pragmatic, and with the use of the Real Options Super Lattice Solver software, even very difficult problems can be easily solved, as will become evident in the next few chapters. This book and software have helped bring the theoretical a lot closer to practice. Firms are using it and universities are teaching it. It is only a matter of time before real options analysis becomes part of normal financial analysis.
- *Real options analysis is just another way to bump up and incorrectly increase the value of a project to get it justified.* Again, nothing is further from the truth. If a project has significant strategic options but the analyst does not value them appropriately, he or she is leaving money on the table. In fact, the analyst will be incorrectly undervaluing the project or asset. Also, one of the foregoing requirements states that one should never run real options analysis unless strategic options and flexibility exist. If they do not exist, then the option value is zero, but if they do exist, neglecting their valuation will grossly and significantly underestimate the project or asset's value.
- *Real options analysis ends up choosing the highest-risk projects as the higher the volatility, the higher the option value.* This criticism is also incorrect. The option value is zero if no options exist. However, if a project is highly risky and has high volatility, then real options analysis becomes more important. That is, if a project is strategic but is risky, then you better incorporate, create, integrate, or obtain strategic real options to reduce and hedge the downside risk and take advantage of the upside uncertainties. Therefore, this argument is actually heading in the wrong direction. It is not that real options will overinflate a project's value, but for risky projects, you should create or obtain real options to reduce the risk and increase the upside, thereby increasing the total strategic value of the project. Also, although an option value is always greater than or equal to zero, sometimes the cost to obtain certain options may exceed their benefit, making the entire strategic value of such options negative, although the option value itself is always zero or positive.

So, it is incorrect to say that real options will always increase the value of a project or that only risky projects are selected. People who make these criticisms do not truly understand how real options work. However, having said that, real options analysis is just another financial analysis tool, and the old axiom of "garbage in, garbage out" still holds. But if care and due diligence are exercised, the analytical process and results can provide highly valuable insights. In fact, I believe that 50 percent (rounded, of course) of the challenge and value of real options analysis is simply *thinking about it*. Understanding that you have options, or obtaining options to hedge the risks and take advantage of the upside, and to think in terms of strategic options, is half the battle. Another 25 percent of the value comes from actually running the analysis and obtaining the results. The final 25 percent of the value comes from being able to explain it to management, to your clients, and to yourself, such that the results become actionable, and not merely another set of numbers. See Chapter 15 for getting started examples using real options software and additional resources for case studies, theoretical concepts, and other applications.

3 Origins of Real Options in Engineering

Harriet Black Nembhard
Pennsylvania State University

Mehmet Aktan
Atatürk University

CONTENTS

As discussed in Chapter 1, the key to real options—from an engineering point of view—is flexibility. In this chapter, we briefly discuss the mathematical and strategic planning research on flexibility that laid the foundations for research in real options.

3.1 MATHEMATICAL FORMULATIONS FOR FLEXIBILITY

From the mid-1970s, math programming approaches to model flexibility began to surge. Salmi (1975) developed a stochastic programming formulation to deal with a single-product, two-country, two-time-period, production-switching problem. The probabilistic nature of exchange rates is captured by discrete probability estimates. The firm maximizes global yield subject to a set of constraints on resource capacities, transfer prices, company interest rates, local loans, intercompany loans, and forward contracts. Pomper (1976) developed a deterministic multiperiod dynamic programming formulation for an international resource allocation and investment problem. The study includes strategic decisions (i.e., choice of location, technology, capacity, and time phasing of new facilities) and also tactical decisions that determine the sourcing pattern of the firm. In this approach, exchange rate scenarios are specified by the decision maker.

Hodder (1982) developed a single-period mixed-integer programming formulation that combines decisions for plant location, resource allocation, and local borrowing. The model maximizes the total profit in terms of the home country's currency for a multinational firm. This approach represents the first attempt to jointly optimize three types of corporate decisions (plant location, resource allocation, and borrowing) in an international context. Jucker and Carlson (1976) proposed a plant location model that permits uncertainty in either price or demand. The firm maximizes its expected utility for end-of-period profits by adopting a mean variance objective function. In this approach, all random variables are assumed to be uncorrelated. Linearity in the supply cost function permits the decomposition of the problem, which can then be solved by standard branch-and-bound techniques.

Hodder (1984) proposed a financial markets approach to the uncapacitated facility location problem under uncertainty. Prices and cost parameters are assumed to be random. The effect of uncertainty in a single-period model was addressed using a mean variance objective function by Hodder and Jucker (1985a, 1985b) and Hodder and Dincer (1986). Hodder and Jucker (1985a, 1985b) solved a stochastic version of the uncapacitated plant location problem. A single-factor approach was adopted for modeling uncertainty in prices and exchange rates. The resulting mixed-integer, quadratic programming problem can be decomposed and solved for a reasonably large number of potential markets and plant sites using standard branch-and-bound techniques. Hodder and Dincer (1986) presented a model for simultaneously determining international plant locations and financing decisions under uncertainty. The model accounts for correlation between price movements in international markets and subsidized financing. As in Hodder and Jucker (1985a, 1985b), the model optimizes over only a single period, and the supply chain includes only plants and market regions.

3.2 MANUFACTURING FLEXIBILITY

The first investigation of manufacturing flexibility as a primary dimension of competitive strategy of a manufacturing business that we found was given by Hayes and Wheelwright (1984). Since their work, there has been substantial growth in the amount of research on this topic.

One of the most widely recognized typologies for classifying the different dimensions of manufacturing flexibility was developed by Browne et al. (1984). In their original framework, they identified eight distinct types of manufacturing flexibility. Sethi and Sethi (1990) enhanced this framework to include eleven distinct dimensions. These dimensions were machine, material handling, operations, process, routing, product, volume, expansion, program, production, and market. Vokurka and O'Leary-Kelly (2000) identified four additional flexibility dimensions: automation, labor, new design, and delivery. They presented a comprehensive contingency-based framework for examining content-related issues involving the relationships and variables included in past manufacturing flexibility studies. They examined several important research design/methodology issues, such as sampling, data collection, and measurement, and proposed solutions to some identified problems.

Models by Gerwin (1987), Slack (1988), and Parthasarthy and Sethi (1993) suggest that manufacturing flexibility is dependent on the nature of a firm's internal operations and external environment. Gerwin (1987) noted that specific sources of environmental uncertainty require the adoption of certain forms of manufacturing flexibility. Slack's (1988) hierarchical model of flexibility depicts manufacturing flexibility as being contingent on the type of competitive strategy adopted by a firm. His framework implies that different competitive strategies require different forms of manufacturing flexibility in order to improve the firm's competitive performance. Parthasarthy and Sethi (1993) propose a much broader view for the influence of manufacturing flexibility on a firm's performance. Their framework incorporates contingent relationships involving an industry's technological environment and its strategic and organizational structural choices. This framework focuses on only a single dimension of manufacturing flexibility. Vokurka and O'Leary-Kelly (2000) express the four general areas that influence manufacturing flexibility. These are strategy, environmental factors, organizational attributes, and technology.

Gerwin (1993) proposed a strategic perspective that involves the influence of environmental uncertainty on manufacturing flexibility. His model suggests that a firm's level of performance is contingent on its ability to match the appropriate type of flexibility with the type of environmental uncertainty faced by the firm. Suarez et al. (1995) focused on the relationship between market uncertainty and manufacturing flexibility. They modeled the firm performance as a logical fit between manufacturing flexibility and the market environment.

Several studies have found direct relationships between various exogenous variables and manufacturing flexibility. Swamidass and Newell (1987) investigated the effects of environmental uncertainty (including uncertainty about users of a firm's products, competition for raw materials, and government regulations) on manufacturing flexibility. They found support for the hypothesized relationship that environmental uncertainty would have a positive impact on production flexibility. Ward et al. (1995) examined the direct effects on market flexibility with regard to four environmental variables: environmental dynamism, business cost, labor availability, and competitive hostility. The results of their study indicated that only unpredictable changes in the environment (i.e., environmental dynamism) were significantly related to market flexibility.

3.3 MANUFACTURING STRATEGY

Global manufacturing strategy planning models started to emerge in the mid- to late-1980s. Several key papers are discussed here, but the reader should note that they do not use the real options approach to exploit flexibility.

Global manufacturing strategy planning models can be grouped into two fundamental types: network flow models and option valuation models. Network flow models exploit primarily portfolio effects within the firm's global supply chain network. In general, network structure decisions are numerous, but are exercised rather infrequently (e.g., on a periodic basis). Alternatively, option valuation models focus primarily on production switching or sourcing decisions contingent on future states of nature. In general, production options are limited, but can be exercised frequently (e.g., on a continuous basis). Polarization in research has arisen due to the analytical complexities of each modeling approach, that is, network complexity in the first case, and stochastic complexity in the second case (Huchzermeier and Cohen 1996).

Starr (1984) introduced the notion of global network models consisting of suppliers, multiple production stages, and final distribution. The objective of his study was to determine optimal production patterns for different network configurations. Armistead (1987) examined the role of manufacturing strategy within an international network of factories. He proposed alternative strategy options as well as alternative performance measures.

Cohen and Lee (1989) provided a detailed analysis of manufacturing strategy options. They proposed manufacturing policy options for each stage of the supply chain for a manufacturer of personal computers. The supply strategies evaluated were central control, regional control, consolidated suppliers, and multiple sourcing. Plant strategies were described as local market oriented, centralized production, product focused, process focused, and vertically integrated. The distribution strategies considered included consolidated distribution centers, co-location of plants and distribution centers, and co-location of distribution centers and market regions.

Miller et al. (1989) compared the manufacturing capabilities of firms in the United States, Japan, and Europe with respect to price competition, product design flexibility, quality, high product performance, product delivery speed, and on-time delivery. According to their study, Japanese and European firms were less able to compete on price than American firms. However, this shift in international competitiveness was caused by dramatic movements of exchange rates, not by successful manufacturing programs (Huchzermeier 1991).

Flaherty (1986) has described global sourcing policies as the standardization of manufacturing across geographically dispersed plants. Carter and Vickery (1988, 1989) illustrated the effectiveness of various strategies for managing volatile exchange rates in global sourcing, through a series of numerical examples. Carter and Narasimhan (1995) discussed the emergence of global purchasing as a strategic weapon for U.S. firms, and provided guidelines for managers engaged in global sourcing.

Cohen et al. (1989) formulated a normative model of certain key global manufacturing and distribution decisions. Their model is a mixed-integer program concerned with the operation of a network of raw material vendors, plants, and markets. They assumed that the firm has specified its product mix, production technology, plant location, capacity, and lower and upper limits on the amount of each product to be supplied to each market. The decisions considered in the model are which vendors to use, the amount of each product to be produced at each plant, the amount of each product to be supplied to each market, and the optimal flows of raw materials from vendors to plants and of finished goods from plants to markets. The model has multiple periods, and can consider a planning horizon of one or more years. The objective of the model is to maximize after-tax profits in the currency of the firm subject to constraints about material flow, plant capacity, market penetration strategies, and local content rules.

Porter (1990) provided a basic framework for the classification of international business strategies for each activity in a value-added supply chain. Primary activities of the value-added supply chain include supply, production, distribution, marketing, and service.

Ettlie and Penner-Hahn (1994) looked at the degree of focus of a plant's manufacturing strategy and its association to both product and production flexibility. They found that the more focused the firm's manufacturing strategy, the lower its production flexibility, as measured by the number of unique parts scheduled for production throughout the year.

3.4   MERGING OPTIMIZATION AND STRATEGY

From the mid-1980s, we start to see a merging of mathematical optimization and strategy planning problems using real options techniques.

A multiperiod stochastic model explicitly incorporating the option valuation of production shifting in a network was qualitatively described in Kogut (1983, 1985). De Meza and van der Ploeg (1987) provided a one-period stochastic model of production shifting to capture the value of flexibility under uncertainty.

Kogut and Kulatilaka (1989) explicitly valued the option of shifting production between two plants located in two different countries as the exchange rate moves. They showed that although each plant evaluated individually may have a negative net present value, when the two locations are evaluated jointly the value may be positive. This is due to the extra value provided by the option to switch production between plants due to changes in the exchange rate. The firm can exercise its option to switch production in every period. Each option contains the option to switch back. This leads to a series of options. Kogut and Kulatilaka (1994) developed a stochastic dynamic programming formulation to determine the value of production switching between two plants located in different countries as the exchange rate fluctuates. In any period, only one plant produces the required quantities to meet demand. A Markov chain model is developed to determine the option value of maintaining two plants with excess capacity. However, this model becomes analytically intractable for more than one exchange rate process or two production location decisions.

Kulatilaka (1993) presented a simple model of flexibility that can be used to obtain the option value to switch between different modes of operation. He then applied that model to evaluate the incremental cost saving of a dual-fuel boiler over the better of two single-fuel boilers. He found that the value of flexibility exceeds the incremental investment cost of purchasing a dual-fuel boiler. Hsu (1998) illustrated in his work that natural gas power plant owners should view their assets as a series of spark spread call options.

Kulatilaka and Trigeorgis (1994) presented an analysis of the generic flexibility to switch between alternative technologies or operating modes. They showed that if there are switching costs, the resulting decision rule reflects a persistence or hysteresis effect where, even though immediate switching may seem attractive, it may be long-term optimal to wait. They considered options to defer investment, expand or contract production, temporarily shut down and restart operations, abandon for salvage, and default during construction.

Huchzermeier and Cohen (1996) developed a stochastic dynamic programming formulation for the valuation of global manufacturing strategy options with switching costs. Exchange rates are modeled as stochastic diffusion processes that exhibit intercountry correlation. Supply chain network options determine the firm's manufacturing flexibility through production capacity and supply chain network linkages. Overall, the firm maximizes its expected, discounted, global, after-tax value through the exercise of product and supply chain network options, and/or through the exploitation of operational flexibility, contingent on exchange rate realizations. They proposed a multinomial approximation of correlated exchange rate processes.

Dasu and Li (1997) studied the structure of optimal policies for a firm operating plants in different countries, where the relative costs of production between the plants vary over time due to exchange rates. They determined the structure of the optimal policies for when and by how much to alter the production quantities in each plant.

Kouvelis (1999) analyzed global sourcing strategies as operational hedging mechanisms for responding to fluctuating exchange rates. He explored the advantages of operational hedging through global supplier selection, where the firm needs to produce a product from various vendors, some of whom are located in foreign countries. The firm has to decide which vendors to choose, and the quantity that should be purchased from each vendor. The vendor choice and sourced quantity can be adjusted dynamically at the expense of switching costs, or within the constraints of a given supply contract, in response to fluctuating exchange rates. The firm wants to develop a global sourcing strategy that will minimize its total expected purchasing cost. However, as in Kogut and Kulatilaka (1994), closed-form solutions are not attainable for more than one exchange rate process (Aktan, 2003).

Descriptions of specific applications of the theory to various industry decision problems include Paddock et al. (1988), Kogut (1991), Grenadier (1995), Schwartz and Moon (2000), Stonier (1999), Jiao et al. (2006), Levaggi and Moretto (2008), Amram et al. (2006), Chen et al. (2007), Tauer (2006), Fujita (2007), Rocha et al. (2006), and Shockley (2007).

REFERENCES

Aktan, M. 2003. Real options valuation of flexibility in manufacturing and quality. Ph.D. thesis, Department of Industrial Engineering, University of Wisconsin–Madison.

Amram, M., Li, F., and Perkins, C. A. 2006. How Kimberly-Clark uses real options. *Journal of Applied Corporate Finance*. 18(2): 40–47.

Armistead, C. G. 1987. International factory networks. *Issues on International Manufacturing (INSEAD)*, September 1987.

Browne, J., Dubois, D., Rathmill, K., Sethi, S. P., and Stecke, K. E. 1984. Classification of flexible manufacturing systems. *The FMS Magazine* 2(2): 114–117.

Carter, J. R., and Narasimhan, R. 1995. *Purchasing and supply management: Future directions and trends*. Tempe, AZ: Center for Advanced Purchasing Studies.

Carter, J. R., and Vickery, S. K. 1988. Managing volatile exchange rates in international purchasing. *Journal of Purchasing and Materials Management* (Winter): 13–20.

Carter, J. R., and Vickery, S. K. 1989. Currency exchange rates: Their impact on global sourcing. *Journal of Purchasing and Materials Management* (Fall): 19–25.

Chen, C., Chiang, Y., and Chien, C. 2007. Real option analysis for capacity investment planning for semiconductor manufacturing. *International Symposium on Semiconductor Manufacturing*. ISSM Paper: MS-P-126.

Cohen, M. A., Fisher, M., and Jaikumar, R. 1989. International manufacturing and distribution networks: A normative model framework. *Managing International Manufacturing* 13:67–93.

Cohen, M. A., and Lee, H. L. 1989. Resource deployment analysis of global manufacturing and distribution networks. *Journal of Manufacturing and Operations Management* 2:81–104.

Dasu, S., and Li, L. 1997. Optimal operating policies in the presence of exchange rate variability. *Management Science* 43(5): 705–722.

De Meza, D., and van der Ploeg, F. 1987. Production flexibility as a motive for multinationality. *Journal of Industrial Economy* 35:343–352.

Ettlie, J. E., and Penner-Hahn, J, D. 1994. Flexibility ratios and manufacturing strategy. *Management Science* 40(11): 1444–1454.

Flaherty, T. 1986. Coordinating international manufacturing and technology. In *Competition in global industries,* ed. M. E. Porter. Boston: Harvard Business School Press.

Fujita, Y. 2007. Toward a new modeling of international economics: An attempt to reformulate an international trade model based on real option theory. *Physica A* 383:507–512.

Gerwin, D. 1987. An agenda for research on the flexibility of manufacturing processes. *International Journal of Operations and Production Management* 7(1): 38–49.

Gerwin, D. 1993. Manufacturing flexibility: A strategic perspective. *Management Science* 39(4): 395–410.

Grenadier, S. R. 1995. Valuing lease contracts: A real-options approach. *Journal of Financial Economics* 38(3): 297–331.

Hayes, R. H., and Wheelwright, S. C. 1984. *Restoring our competitive edge: Competing through manufacturing.* New York: Wiley.

Hodder, J. E. 1982. Plant location modeling for the multinational firm. In *Proceedings of the Academy of International Business Conference on the Asia-Pacific Dimension of International Business,* Honolulu.

Hodder, J. E. 1984. Financial market approaches to facility location under uncertainty. *Operations Research* 32:1374–1380.

Hodder, J. E., and Dincer, M. C. 1986. A multifactor model for international plant location and financing under uncertainty. *Computers and Operations Research* 13:601–609.

Hodder, J. E., and Jucker, J. V. 1985a. International plant location under price and exchange rate uncertainty. *Engineering and Production Economic* 9:225–229.

Hodder, J. E., and Jucker, J. V. 1985b. A simple plant-location model for quantity-setting firms subject to price uncertainty. *European Journal of Operational Research* 21:39–46.

Hsu, M. 1998. Spark spread options are hot! *The Electricity Journal* 11:28–39

Huchzermeier, A. 1991. Global manufacturing strategy planning under exchange rate uncertainty. Ph.D. thesis, Decision Sciences Department, Wharton School, University of Pennsylvania.

Huchzermeier, A., and Cohen, M. A. 1996. Valuing operational flexibility under exchange rate risk. *Operations Research* 44(1): 100–113.

Jiao, J., Kumar, A., and Lim, C. M. 2006. Flexibility valuation of product family architecture: A real-option approach. *International Journal of Advanced Manufacturing Technology* 30:1–9.

Jucker, J. V., and Carlson, R. C. 1976. The simple plant-location problem under uncertainty. *Operations Research* 24(6): 1045–1055.

Kogut, B. 1983. Foreign direct investment as a sequential process. In *The multinational corporations in the 1980s,* ed. C. P. Kindelberger and D. Audretsch. Cambridge, MA: MIT Press.

Kogut, B. 1985. Designing global strategies: Profiting from operational flexibility. *Sloan Management Review* 27(1): 27–38.

Kogut, B. 1991. Joint ventures and the option to expand and acquire. *Management Science* 37(1): 19–33.

Kogut, B., and Kulatilaka, N. 1989. Foreign direct investment and exchange rate volatility. Working Paper. Philadelphia: Wharton School, University of Pennsylvania.

Kogut, B., and Kulatilaka, N. 1994. Operating flexibility, global manufacturing, and the option value of a multinational network. *Management Science* 40(1): 123–139.

Kouvelis, P. 1999. Global sourcing strategies under exchange rate uncertainty. In *Quantitative models for supply chain management,* ed. S. Tayur, R. Ganeshan, and M. Magazine, pp. 625–667. Boston: Kluwer Academic.

Kulatilaka, N. 1993. The value of flexibility: The case of a dual-fuel industrial steam boiler. *Financial Management* 22(3): 271–280.

Kulatilaka, N., and Trigeorgis, L. 1994. The general flexibility to switch: Real options revisited. *The International Journal of Finance* 6(2): 778–798.

Levaggi, R., and Moretto, M. 2008. Investment in hospital care technology under different purchasing rules: A real option approach. *Bulletin of Economic Research* 60(2): 159–181.

Miller, J. G., Amano, A., De Meyer, A., Ferdows, K., Nakane, J., and Roth, A. 1989. Closing the competitive gaps: The international report of the manufacturing futures survey. *Managing International Manufacturing* 13:153–168.

Paddock, J. L., Siegel, D. R., and Smith, J. L. 1988. Option valuation of claims on real assets: The case of offshore petroleum leases. *Quarterly Journal of Economics* 103(3): 479–508.

Parthasarthy, R., and Sethi, S. P. 1993. Relating strategy and structure to flexible automation: A test of fit and performance implications. *Strategic Management Journal* 14(7): 529–549.

Pomper, C. L. 1976. *International investment planning: An integrated approach.* Amsterdam: North-Holland.

Porter, M. E. 1990. The competitive advantage of nations. *Harvard Business Review* 68(2): 73–93.

Rocha, K., Moreira, A, Reis, E. J., and Carvalho, L. 2006. The market value of forest concessions in the Brazilian Amazon: A real option approach. *Forest Policy and Economics* 8:149–160.

Salmi, T. 1975. *Joint determination of trade, production, and financial flows in the multinational firm assuming risky currency exchange rates.* Helsinki: Helsinki School of Economics.

Schwartz, E., and Moon, M. 2000. Rational pricing of Internet companies. *Financial Analysts Journal* 56:62–75.

Sethi, A. K., and Sethi, S. P. 1990. Flexibility manufacturing: A survey. *International Journal of Flexible Manufacturing Systems* 2(4): 289–328.

Shockley, R. L. 2007. A real option in a jet engine maintenance contract. *Journal of Applied Corporate Finance* 19(2): 88–94.

Slack, N. 1988. Manufacturing systems flexibility: An assessment procedure. *Computer Integrated Manufacturing Systems* 1(1): 25–31.

Starr, M. K. 1984. Global production and operations strategy. *Columbia Journal of World Business* 19(4): 17–22.

Stonier, J. E. 1999. What is an aircraft purchase option worth? Quantifying asset flexibility created through manufacturer lead-time reductions and products commonality. In *Handbook of Airline Finance,* ed. Aviation Week Group, pp. 231–250. Washington, DC: Aviation Week Group.

Suarez, F. F., Cusumano, M. A., and Fine, C. H. 1995. An empirical study of flexibility in manufacturing. *Sloan Management Review* 37(1): 25–32.

Swamidass, P. M., and Newell, W. T. 1987. Manufacturing strategy, environmental uncertainty and performance: A path analytic model. *Management Science* 33(4): 509–524.

Tauer, L. W. 2006. When to get in and out of dairy farming: A real option analysis. *Agricultural and Resource Economics Review* 35(2): 339–347.

Vokurka, R. J., and O'Leary-Kelly, S. W. 2000. A review of empirical research on manufacturing flexibility. *Journal of Operations Management* 18:485–501.

Ward, P. T., Duray, R., Leong, G. K., and Sum, C. 1995. Business environment, operations strategy, and performance: An empirical study of Singapore manufacturers. *Journal of Operations Management* 13(2): 99–115.

4 Real Options in Manufacturing Operations

Harriet Black Nembhard
Pennsylvania State University

Mehmet Aktan
Atatürk University

Leyuan Shi
University of Wisconsin–Madison

CONTENTS

By using the real options approach, manufacturing firms can value the flexibility that exists or that is planned for operation. In addition, firms can gain insight on the optimal strategic decisions for future price levels. Real options valuation also enables sensitivity analysis of its inputs, that is, impacts of input parameter changes on profitability can be estimated. Several scenarios show that real options analysis can be used to support managerial decision making for manufacturing-related operations, such as contracting a portion of the production, abandoning the operations, choosing among a number of strategies, and switching technologies.

4.1 INTRODUCTION

A firm may have the right to delay, expand, contract, switch, or abandon a manufacturing project for a given cost or salvage value at some future date, which are real options. The value of the project may be increased if it is optimal to exercise the project's embedded option. Whether or not it is optimal depends on market conditions. The option need not be exercised if conditions are not favorable.

Thinking about investment projects in option terms encourages managers to decompose an investment into its component options and risks, which can lead to valuable insights about sources of uncertainty and how uncertainty will resolve over time (Brabazon 1999). Options thinking also encourages managers to consider how to enhance the value of their investments by building in more flexibility where possible. Real options analysis has the potential to allow companies to examine programs of capital expenditures as multiyear investments, rather than as individual projects (Copeland 2001).

If the analogous real options parameters can be estimated, any method used to value financial options can potentially be used to value real options. Charnes (2007), Amram and Kulatilaka (1999), Copeland and Antikarov (2001), and Mun (2002, 2003) provide guidelines for analyzing real options.

In this chapter, sample real options applications in manufacturing operations are presented. Real option types related with manufacturing operations that will be presented are contract, abandon, choose, and switch options. Valuations of the options are demonstrated using binomial lattices.

4.2   BINOMIAL LATTICES

A binomial lattice is a tree that represents different possible paths that might be followed by the state variable over the life of the option. Suppose that the initial value of the state variable is S, and the state variable either increases or decreases at each predefined time interval. The rate of an up move for the state variable is denoted by u, and the rate of a down move is denoted by d for each time interval, where $u > 1 > d > 0$. At the end of an interval with a time length of Δt, the value of the state variable will be either uS with probability p, or dS with probability $1 - p$. The value of a single-period European call option in a risk-neutral environment is approximated by

$$c = \left(pc_u + (1-p)c_d \right)e^{-r\Delta t} = \left(\frac{e^{r\Delta t} - d}{u - d}c_u + \frac{u - e^{r\Delta t}}{u - d}c_d \right)e^{-r\Delta t} \tag{4.1}$$

where c_u is the value of the call option if the state variable value is uS at the end of the time interval, and c_d is the value of the call option if the state variable value is dS at the end of the time interval. The u and d values are calculated using the volatility, σ, for the state variable as follows:

$$u = e^{\sigma\sqrt{\Delta t}} \quad \text{and} \quad d = 1/u \tag{4.2}$$

Since value of money can change with time, a reference point must be selected to make financial evaluations. Because the interest rate is the more identifiable and accepted measure of the earning power of money, it is usually used as a parameter in the analysis for time value of money. In the net present value (NPV) method, the present time (time zero) equivalent value of all the costs and benefits incurred during the life of a system or a project is calculated using a specific interest rate, which is denoted by r in Equation 4.1. The NPV of a cash amount C that will be gained after a time amount of Δt years is given by

$$NPV = Ce^{-r\Delta t}$$

where r is continuously compounded yearly interest rate. The NPV calculation in Equation 4.1 is also made using continuous compounding. In the following sections, real options valuation using multiple-step binomial lattices will be presented.

4.3   OPTION TO CONTRACT

Suppose that an aeronautical manufacturing firm is unsure of the technological efficacy and market demand of its new fleet of long-range supersonic jets. The firm decides to hedge itself through the use of strategic options, specifically an option to contract 70 percent of its manufacturing facilities at any time within the next five years. Suppose that the present value of the expected future cash flows discounted at an appropriate market risk-adjusted discount rate is $S = \$1$ billion for the firm's current operating structure. Volatility of the logarithmic returns on the projected future cash flows is $\sigma = 0.20$. The risk-free rate on a riskless asset for the next five years is found to be yielding $r = 3$ percent. Suppose the firm has the option to contract 70 percent of its current operations at any time over the next five years, thereby creating an additional $400 million in savings after this contraction (Mun 2003).

Tables 4.1 through 4.3 show step-by-step analysis using a binomial approach to solve the contraction option, which is estimated to be $101.93 million using a five-time-step lattice. Since the asset-pricing lattice is presented as a spreadsheet, moving one column to the right in a row means a value increase with the rate of u, and moving one column to the right and one row down together means a value decrease with the rate of d as shown in Table 4.1. The value of contracting 70 percent of its operations is equivalent to 70 percent of its existing operations plus the $400 million in savings.

TABLE 4.1
Asset Pricing Lattice

1000.00	1221.40	1491.82	1822.12	2225.54	2718.28
	818.73	1000.00	1221.40	1491.82	1822.12
		670.32	818.73	1000.00	1221.40
			548.81	670.32	818.73
				449.33	548.81
					367.88

Note: Time step size $(dt) = 1$. Up jump size $(u) = e^{\sigma\sqrt{dt}} = 1.2214$. Down jump size $(d) = 1/u = 0.8187$. Risk-neutral probability $(p) = \frac{e^r - d}{u - d} = 0.5258$.

TABLE 4.2

Option Valuation Lattice

1101.93	1281.93	1520.26	1829.23	2225.54	2718.28
	973.11	1100.00	1275.31	1507.28	1822.12
		869.22	973.11	1100.00	1254.98
			784.17	869.22	973.11
				714.53	784.17
					657.52

TABLE 4.3

Decision Lattice

continue	continue	continue	continue	continue	end
	contract	contract	continue	continue	end
		contract	contract	contract	contract
			contract	contract	contract
				contract	contract
					contract

Management will choose the strategy that maximizes profitability. The calculations are made moving backward, beginning from the last column in the asset price lattice. Maximum profit in a node of the asset price lattice can be found as

$$\max_{i,j} \underbrace{[(pS_{(i+1,j)} + (1-p)S_{(i+1,j+1)})e^{-r}}_{\text{"continue"}}, \underbrace{0.7S + \$400 \text{ million}]}_{\text{"contract"}} \qquad (4.3)$$

decisions

where i is the column number and j is the row number of the node, and $S_{i,j}$ is the asset price in node$_{i,j}$. The profit-maximizing decision is to "continue" if the value of the first element in brackets is larger than the second element, and it is to "contract" if the value of the second element in brackets is larger than the first element in Equation 4.3.

The value of continuing is simply the discounted weighted average of potential future option values using the risk-neutral probability. Using the backward induction technique, the lattice is back-calculated to the starting point to obtain the value of $1,101.93 million. Because the NPV of the expected future cash flows discounted at an appropriate market risk-adjusted discount rate is $S = \$1$ billion for the firm's current operating structure, the option value of being able to contract 70 percent of its operations is $101.93 million. $1,000 million is the static NPV without flexibility, $101.93 million is the real options value, and the combined value of $1,101.93 million is the expanded NPV (ENPV), or NPV with real options flexibility value, the correct total value of this manufacturing initiative. The real options value is worth an additional 10.193 percent of existing business operations. If a real options approach is not used, the manufacturing initiative will be undervalued (Mun 2003).

4.4 OPTION TO ABANDON

Let us consider the manufacturing company scenario given in Section 4.3. Due to the uncertain nature of technological efficacy and market demand, management decides that it will create a strategic abandonment option. That is, at any time period within the next five years, management can review the progress of the research and development effort and decide whether to terminate the research for its long-range supersonic jets. After five years, the firm would have either succeeded or completely failed in its supersonic jets development initiative, and there exists no option value after that time period. If the program is terminated, the firm can potentially sell its production rights to another aeronautical firm with which it has a contractual agreement. This contract with the other firm is exercisable at any time within this period (Mun 2003).

As in Section 4.3, suppose that the present value of the expected future cash flows is $S = \$1$ billion, the volatility of the logarithmic returns on the cash flows is $\sigma = 0.20$, and the risk-free rate is $r = 3$ percent. The value of the production rights is $900 million contractually, if sold within the next five years.

Tables 4.4 through 4.6 show the analysis of the binomial approach to value the abandonment option, which is estimated to be $81.79 million using a five-time-step lattice. The value of abandonment is $900 million.

TABLE 4.4
Asset Pricing Lattice

1000.00	1221.40	1491.82	1822.12	2225.54	2718.28
	818.73	1000.00	1221.40	1491.82	1822.12
		670.32	818.73	1000.00	1221.40
			548.81	670.32	818.73
				449.33	548.81
					367.88

Note: Time step size $(dt) = 1$. Up jump size $(u) = e^{\sigma\sqrt{dt}} = 1.2214$. Down jump size $(d) = 1/u = 0.8187$. Risk-neutral probability $(p) = \frac{e^r - d}{u - d} = 0.5258$.

TABLE 4.5
Option Valuation Lattice

1081.79	1255.91	1499.74	1822.12	2225.54	2718.28
	958.21	1066.20	1238.61	1491.82	1822.12
		900.00	943.51	1037.40	1221.40
			900.00	900.00	900.00
				900.00	900.00
					900.00

TABLE 4.6

Decision Lattice

continue	continue	continue	continue	continue	end
	continue	continue	continue	continue	end
		abandon	continue	continue	end
			abandon	**abandon**	**abandon**
				abandon	**abandon**
					abandon

Management will choose the strategy that maximizes profitability. The calculations are made moving backward, beginning from the last column in the asset price lattice. Maximum profit in a node of the asset price lattice can be calculated as

$$\underset{i,j}{\underbrace{\max}_{\text{decisions}}} [\underbrace{(pS_{(i+1,j)} + (1-p)S_{(i+1,j+1)})e^{-r}}_{\text{"continue"}}, \underbrace{\$900 \text{ million}}_{\text{"abandon"}}] \tag{4.4}$$

$\Longrightarrow$

where i is the column number and j is the row number of the node, and $S_{i,j}$ is the asset price in node$_{i,j}$. The profit-maximizing decision is to "continue" if the value of the first element in brackets is larger than $900 million, and it is to "abandon" if $900 million is larger than the first element in Equation 4.4.

The lattice is back-calculated to the starting point to obtain the value of $1,081.79 million. Because the value of the operations without the option to abandon is $1,000 million, the option value of being able to abandon its operations is $81.79 million. The static NPV without flexibility is $1,000 million, and the real options value is $81.79 million. The combined value of $1,081.79 million is the ENPV. The real options value is worth an additional 8.179 percent of existing operations (Mun 2003).

4.5   OPTION TO CHOOSE

Suppose the manufacturing firm represented in Sections 4.3 and 4.4 decides to hedge itself through the use of strategic options. Specifically, it has the option to choose among two strategies: (1) contracting its manufacturing operations, or (2) completely abandoning its business unit at any time within the next five years, that is, the two strategies discussed in Sections 4.3 and 4.4. Let us suppose again that the firm has a current operating structure whose static valuation of future profitability is $S = \$1,000$ million. Volatility of the logarithmic returns on cash flows is $\sigma = 0.20$. The risk-free rate on a riskless asset for the next five years is 3 percent annually. The firm has the option to contract 70 percent of its current operations at any time over the next five years, thereby creating an additional $400 million in savings after this contraction, as in Section 4.3. By abandoning its operations, the firm can sell its intellectual property for $900 million, as in Section 4.4.

If the project is analyzed separately, we get differing and misleading valuations as follows: the contraction option only is worth $101.93 million, and the abandonment option only is worth $81.79 million. The sum of the two individual option values is $183.72 million. Clearly, valuing a combination of real options by performing them individually and then summing them yields incorrect results. The reason why the sum of individual options does not equal the interaction of the same options is due to the mutually exclusive and independent nature of these specific options. That is, the firm cannot both contract and abandon on the same node at the same time. This mutually exclusive behavior is captured using the option to choose (contract or abandon). If performed separately on a particular node in the lattice, the contraction option analysis may indicate that it is optimal to contract, while the abandonment option analysis may indicate that it is optimal to abandon, creating a higher total value. However, in an option to choose, the option is not overvalued since multiple option executions cannot occupy the same state (Mun 2003).

Tables 4.7 through 4.9 show the analysis of the binomial approach to value the option to choose, which is estimated to be $106.95 million using a five-time step lattice.

TABLE 4.7
Asset Pricing Lattice

1000.00	1221.40	1491.82	1822.12	2225.54	2718.28
	818.73	1000.00	1221.40	1491.82	1822.12
		670.32	818.73	1000.00	1221.40
			548.81	670.32	818.73
				449.33	548.81
					367.88

Note: Time step size $(dt) = 1$. Up jump size $(u) = e^{\sigma\sqrt{dt}} = 1.2214$. Down jump size $(d) = 1/u = 0.8187$. Risk-neutral probability $(p) = \frac{e^r - d}{u - d} = 0.5258$.

TABLE 4.8
Option Valuation Lattice

1106.95	1282.80	1520.26	1829.23	2225.54	2718.28
	983.05	1101.89	1275.31	1507.28	1822.12
		914.42	980.38	1100.00	1254.98
			900.00	910.71	973.11
				900.00	900.00
					900.00

TABLE 4.9
Decision Lattice

continue	continue	continue	continue	continue	end
	continue	continue	continue	continue	end
		continue	continue	**contract**	**contract**
			abandon	continue	**contract**
				abandon	**abandon**
					abandon

Management will choose the strategy that maximizes profitability. The calculations are made moving backward, beginning from the last column in the asset-pricing lattice. Maximum profit in a node of the asset-pricing lattice can be calculated as

$$\max_{i,j}[\underbrace{(pS_{(i+1,j)}+(1-p)S_{(i+1,j+1)})e^{-r}}_{\text{"continue"}},\ \underbrace{0.7S+\$400\text{ million}}_{\text{"contract"}},\ \underbrace{\$900\text{ million}}_{\text{"abandon"}}] \qquad (4.5)$$

decisions

$\Longrightarrow$

where i is the column number and j is the row number of the node, and $S_{i,j}$ is the asset price in node$_{i,j}$. The profit-maximizing decision is to "continue" if the value of the first element in brackets is larger than the second and third elements (contract and abandon), it is to "contract" if the value of the second element in brackets is larger than the first and third elements (continue and abandon), and it is to "abandon" if $900 million is larger than the values of the first and second elements (continue and contract) in Equation 4.5.

The lattice is back-calculated to the starting node to obtain the value of $1,106.95 million. Since the value without the option to choose is $1,000 million, the value of the option to choose is $106.95 million. The combined value of $1,106.95 million is the ENPV. The option to choose is worth an additional 10.695 percent of existing operations.

Since the option to choose has more flexibility than the individual options of contracting or abandoning alone, its value must be greater than or equal to the most valuable of these two options. As can be seen from the results, the value of option to choose is $106.95 million, and it is greater than both the contracting and abandonment option values, which are $101.93 million and $81.79 million, respectively.

4.6   SWITCHING OPTION

A switching option in manufacturing operations gives the flexibility of being able to switch elements such as resources, assets, and technology. This ability to switch provides added value to manufacturing operations, but it is often subject to a switching cost.

When switching is costly, a switch decision is affected from the previous decisions. In other words, the decision in a node depends on the strategies and asset prices in the preceding nodes. Therefore, there are usually many possible sets of dependent decisions in the lattice, and displaying all possible decisions on a lattice is not practical.

Assume that a manufacturing firm considers two different technologies in its manufacturing operations. The first technology is worth $1,000, and the second technology is worth $800. Assume that the firm is currently using the first technology, and it can switch to the second technology within five years. The cost of switching to the second technology is 5 percent of the first technology's worth, and the risk-free rate is 3 percent. The first asset price has a volatility of $\sigma = 0.25$, and the second has a volatility of $\sigma = 0.20$. Assume that the correlation coefficient of the two asset prices is −0.30.

In order to value the switching option, optimal decisions on a lattice node depending on each connected previous node's strategy must be found. After this operation is done for all nodes, a set of optimal strategies and maximum expected profit can be obtained. The maximum expected profit resulting from optimal strategies is the value of the switch option. The calculated maximum profit for the given parameters is $182.44. The static NPV of switching is −$250 = $800 − 1.05 × $1,000. Then, the real option value is $432.44 = $182.44 − (−$250). Although it now costs $50 to switch (5 percent of $1,000) from the first technology to the second, which is a less profitable asset now ($800), there is strategic value because of asset price volatility on both technologies. There is a chance that the second asset's value may overtake the first asset's value. In addition, due to the negative correlation coefficient, the ability to switch to the second asset provides a risk diversification effect for the first asset, making the flexibility to switch valuable.

Assume that the correlation coefficient is 0.30 instead of −0.30, and all other parameters are the same. A positive correlation provides less risk diversification and therefore reduces the value of the switching option. The ability to switch is more important if both assets move inversely with each other. That is, when the existing asset value decreases, the option holder will switch to the second asset whose value is increasing due to the negative correlation (Mun 2003). When the correlation coefficient is 0.30 instead of −0.30, the value of the switch option decreases to $365 = $115 − (−$250) from $432.44, as expected.

A more detailed discussion and application of switching options are presented in Chapter 6.

4.7 SUMMARY

This chapter has provided applications of real options models on manufacturing operations. By using the real options approach, manufacturing firms can value the flexibility that exists in the system, or is planned to be added to the system. In addition, firms can gain insight on the optimal strategic decisions for future possible levels of prices. The real options approach reveals the additional value of uncertainty, which cannot be found by static NPV analysis. The calculations were all conducted using Real Options Analysis Toolkit 1.0. Real options valuation also enables sensitivity analysis of its inputs (i.e., impacts of input parameter changes on profitability can be estimated).

The scenarios covered show that real options analysis can be used to support managerial decision making for manufacturing-related operations, such as contracting a portion of the production, abandoning the operations, choosing among a number of strategies, and switching technologies.

REFERENCES

Amram, M., and N. Kulatilaka. 1999. *Real options: Managing strategic investment in an uncertain world*. Boston: Harvard Business School Press.

Brabazon, T. 1999. Real options: Valuing flexibility in capital investment decisions. *Accountancy Ireland*, 31(6) 16–18.

Charnes, J. 2007. *Financial Modeling with Crystal Ball and Excel*. Hoboken, NJ: John Wiley.

Copeland, T. 2001. The real options approach to capital allocation. *Strategic Finance*, 83(4) 33–37.

Copeland, T., and V. Antikarov. 2001. *Real options: A practitioner's guide*. New York, Texere Publishing Limited.

Mun, J. 2002. *Real options analysis: Tools and techniques for valuing strategic investments and decisions*. Hoboken, NJ: John Wiley.

Mun, J. 2003. *Real options analysis course: Business cases and software applications*. Hoboken, NJ: John Wiley.

5 Real Options Valuation for Quality Improvement

Harriet Black Nembhard
Pennsylvania State University

Mehmet Aktan
Atatürk University

CONTENTS

In this chapter, a real options framework is presented to value the quality improvement programs and to determine strategic decisions about the programs. A basic financial model is developed to present the economic benefits of the program. A multinomial lattice technique is used to value the expected benefit potential of the quality improvement program. Sensitivity of the option value against the changes in parameters of the quality improvement program is also investigated.

5.1 QUALITY IMPROVEMENT APPROACHES

Quality has become one of the most important consumer decision factors in the selection among competing products and services. Consequently, understanding and improving quality are key factors leading to business success, growth, and an enhanced competitive position. There is a substantial return on investment from improved quality, and from successfully employing quality as an integral part of overall business strategy. Quality improvement methods can be applied to any area within a company or organization, including manufacturing, process development, engineering design, finance and accounting, marketing, and field service of products.

Over the past approximately ten years, there has been an increasing interest in the use of the Six Sigma methodology for process and product quality improvement. This methodology builds upon statistical tools such as statistical process control (SPC) and design of experiments as well as management structures such as total quality management (TQM) and team-building approaches to improve quality (Breyfogle 2003).

By approaching quality improvement in this way, companies can create or redesign processes, products, and services for top performance. As a result, they can reduce their manufacturing or service costs, and they can increase the number of their customers with higher-quality products and brand reputation (Persse 2006; Papadakis 2007). The economic benefits of higher-quality products include reduced product recall rates and lower product warranty costs for companies.

The real options approach can be used to analyze the potential economic benefit that a project may bring to a company. This approach was used to value the option of implementing quality control charts in a company by Nembhard et al. (2002). By presenting the problem here, we introduce a basic framework for demonstrating the financial model and real options valuation using the well-known lattice structure.

5.2   FINANCIAL MODEL

Let us assume that two variables affect the profit function of a manufacturing company. Let S_t be the sales price of the product during time interval t. Let V_t be the variable production cost for the product during time interval t. Let D be demand per time interval. Then, total revenue during time interval t is

$$R_t = S_t D \tag{5.1}$$

Assuming that demand is equal to the number of units produced in each time interval, profit P_t in time interval t can be defined as

$$P_t = R_t - F - V_t D \tag{5.2}$$

where F is the fixed production cost per time interval and V_t is the variable production cost per unit product.

Let d be the rate of extra demand, and let v be the rate of variable production cost savings if a quality improvement program is implemented. Let K be the cost of implementing a quality improvement program per time interval. The profit per time interval can be represented as

$$P_t = (1 + d)S_t D - F - (1 - v)V_t D - K \rightarrow \text{with quality improvement} \tag{5.3a}$$

or

$$P_t = S_t D - F - V_t D \rightarrow \text{without quality improvement} \tag{5.3b}$$

Then, profit increase due to a quality improvement program during time interval t can be defined as the maximum of the difference between the two profit functions above, and zero:

$$G_t = \max(0, dS_t D + vV_t D - K) \tag{5.4}$$

Other factors such as savings in product warranty costs, reduced recall for products, and service cost savings can easily be included in the model since such savings or increased profits can be represented as a function of the number of products sold or produced.

5.3 REAL OPTIONS VALUATION

Lattice techniques are commonly used numerical valuation approaches for multivariate option models. Let us consider a lattice technique to model and value a quality improvement program.

Since there are two variables in the profit function, the two-state KR approach (Kamrad and Ritchken 1991) can be used to find the option value and the optimal strategy for the quality improvement program. The KR approach proposes a pentanomial lattice structure for modeling two variables. Assume that the joint density of variables S_1 and S_2 is bivariate lognormal. For variable $i(i = 1, 2)$, let the instantaneous mean be $\mu_i = r - \sigma_i^2/2$, where r is the risk-free interest rate, and let the instantaneous variance be σ_i^2. In each time step with length Δt, values of the two variables can increase or decrease, or they can stay stable. An up move for variable i is denoted by u_i, and a down move for variable i is denoted by d_i. There are five paths leaving each node in the pentanomial lattice. Amounts of five jumps for the variables and corresponding jump probabilities are presented in Table 5.1.

Jump probabilities p_1, p_2, p_3, and p_4 are

$$p_1 = \frac{1}{4}\left[\frac{1}{\lambda^2} + \frac{\sqrt{\Delta t}}{\lambda}\left(\frac{\mu_1}{\sigma_1} + \frac{\mu_2}{\sigma_2}\right) + \frac{\rho}{\lambda^2}\right] \tag{5.5a}$$

$$p_2 = \frac{1}{4}\left[\frac{1}{\lambda^2} + \frac{\sqrt{\Delta t}}{\lambda}\left(\frac{\mu_1}{\sigma_1} - \frac{\mu_2}{\sigma_2}\right) - \frac{\rho}{\lambda^2}\right] \tag{5.5b}$$

$$p_3 = \frac{1}{4}\left[\frac{1}{\lambda^2} + \frac{\sqrt{\Delta t}}{\lambda}\left(-\frac{\mu_1}{\sigma_1} - \frac{\mu_2}{\sigma_2}\right) + \frac{\rho}{\lambda^2}\right] \tag{5.5c}$$

$$p_4 = \frac{1}{4}\left[\frac{1}{\lambda^2} + \frac{\sqrt{\Delta t}}{\lambda}\left(-\frac{\mu_1}{\sigma_1} + \frac{\mu_2}{\sigma_2}\right) - \frac{\rho}{\lambda^2}\right] \tag{5.5d}$$

TABLE 5.1

Variable Jump Amounts and Probabilities

S_1	S_2	Probability
u_1	u_2	p_1
u_1	d_2	p_2
d_1	d_2	p_3
d_1	u_2	p_4
1	1	p_5

Note: $u_i = e^{\lambda \sigma_i \sqrt{\Delta t}}$ $(i = 1, 2)$ and $\lambda \geq 1$.

where ρ is the correlation of the logarithmic returns of the two variables.

Since $p_1 + p_2 + p_3 + p_4 + p_5 = 1$, we have

$$p_5 = 1 - \frac{1}{\lambda^2} \tag{5.5e}$$

It is convenient to reduce the number of nodes by imposing the condition that $u_i d_i = 1$, so that an up followed by a down is equal to 1, which means that the value of variable i turns back to the initial level at the end of two time steps. The total number of nodes after n iterations is

$$\sum_{i=10}^{n} 5^i = (5^{n+1} - 1)/4 \quad \text{if} \quad u_i d_i \neq 1$$

and

$$\sum_{i=10}^{n} [(i+1)^2 + i^2] = (4n^3 + 12n^2 + 14n + 6)/6 \quad \text{if} \quad u_i d_i = 1 \tag{5.6}$$

In ten iterations, a pentanomial lattice will have 12,207,031 nodes if $ud \neq 1$, whereas there will be only 891 nodes if we impose $u_i d_i = 1$.

Figure 5.1 shows the first two iterations of a pentanomial lattice with the condition $u_i d_i = 1$. The first element in parentheses shows the change in the first variable, and the second element shows the change in the second variable. As can be seen in Figure 5.1, there are a total of nineteen nodes in the pentanomial lattice for two iterations when $u_i d_i = 1$.

Last-column elements in the lattice are calculated as $G_t = \max(0, dS_tD + vV_tD - K)$ as in Equation 5.4. Backward discounting is applied on the lattice starting from the last time interval. First, an expected value is found by multiplying the jump probabilities (see Equation 5.5) with the corresponding G_t values in the five nodes.

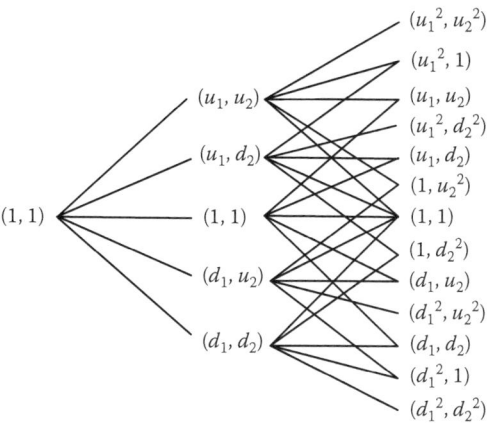

FIGURE 5.1 Pentanomial lattice structure with two time steps.

Next, this expected value is discounted with the risk-free interest rate in one interval. This value is the expected amount of profit for the last step, which is determined from the node in the preceding time interval. The expected discounted value is added to the profit value in the origin node for the five following nodes, so that the expected total profit is found for those two time intervals. When this calculation is done for all nodes in one time interval, and then for all remaining nodes going back one time interval in each iteration, the expected discounted value at time zero is determined. The option value c, which is the expected additional profit gained by optimally implementing the quality improvement project, is equal to this discounted value.

It is assumed that switching between the strategies of "implementing" and "not implementing" the quality improvement program does not impart a cost to the firm. In other words, there is a cost of implementing the quality improvement program per time interval when it is implemented, but there is no additional cost of stopping or restarting the program. This assumption will be relaxed in later chapters, but with this assumption, the option valuation model can be handled as a total of n European options, where n is the number of time steps that the quality improvement program can be implemented. Without this costless switching assumption, each decision affects decisions given in other time steps, and such models can be solved with an appropriate dynamic programming algorithm.

5.4   EXAMPLE OF QUALITY IMPROVEMENT VALUATION

Let us consider an example for the costless switching scenario. Assume that a company is considering the implementation of a quality improvement program within the next three years. Implementation of the program will be reviewed every month, and at the beginning of each month a decision will be given whether to implement the program at that month or not. So, a total of thirty-six decisions will be given during the program's time span of three years. Let us assume that the current price

TABLE 5.2

Jump Probabilities and Jump Amounts

$p_1 =$	0.202	$u_1 =$	1.1095
$p_2 =$	0.139	$u_2 =$	1.1095
$p_3 =$	0.214	$d_1 =$	0.9013
$p_4 =$	0.139	$d_2 =$	0.9013
$p_5 =$	0.306		

of the company's product is $S_0 = \$1,500$, the current variable production cost for the product is $S_0 = \$500$, the demand for the product is $D = 1,000$ units per month, the risk-free interest rate is $r = 3$ percent, the yearly volatility (standard deviation of the logarithmic returns) for the product's price is $\sigma_1 = 0.3$, the yearly volatility for the variable production cost of the product is $\sigma_2 = 0.3$, the correlation coefficient for the logarithmic returns of the two variables is $\rho = 0.2$, the fixed production cost per month is $F = \$300,000$, the rate of extra demand if quality improvement is implemented is $d = 3$ percent, the cost of implementing a quality improvement program per month is $K = \$20,000$, and the rate of variable production cost savings if a quality improvement program is implemented is $v = 2$ percent.

Jump probabilities and jump amounts can be calculated with the information provided above, as shown in Table 5.2.

Using the values above in the pentanomial lattice, the real options value of the quality improvement project is estimated to be $1,327,425. This amount is the expected worth of keeping the quality improvement project as an option for three years. A pentanomial lattice with thirty-six time steps and 33,781 nodes is generated to estimate the option value.

It is instructive to also consider how much the option value is affected by changes in system parameters. The rate of extra demand if quality improvement is implemented (d), the rate of variable production cost savings if a quality improvement program is implemented (v), and the cost of implementing a quality improvement program per time interval (K) are critical parameters that are directly related to the quality improvement project. We can determine their impact on the real options value by running a sensitivity analysis.

Figure 5.2 shows the option value versus the percentage of demand increase that occurs with the implementation of the quality improvement program. Figure 5.3 shows the option value versus the percentage of variable production cost savings that occurs with the implementation of a quality improvement program. Figure 5.4 shows the option value versus the monthly implementation cost of the quality improvement program.

In general, this type of sensitivity analysis may guide the decision maker about which ranges suggest the implementation of a quality improvement option.

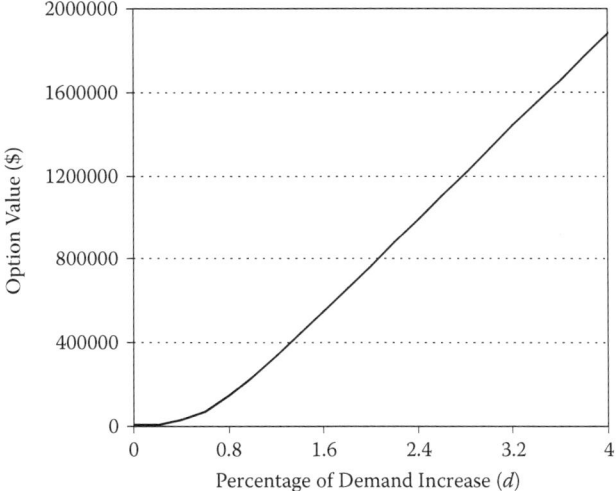

FIGURE 5.2 Option value against d.

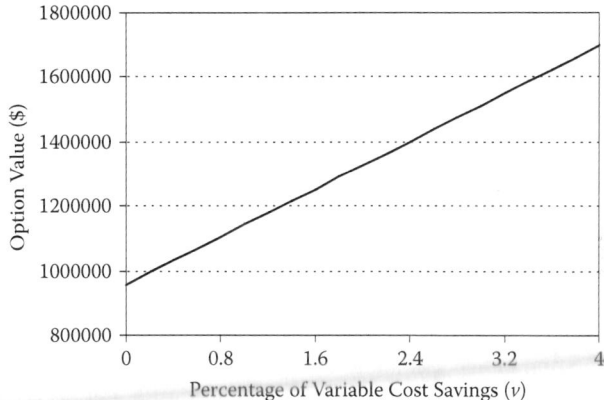

FIGURE 5.3 Option value against v.

Figure 5.2 suggests that the company may derive more benefit with increasing d (rate of extra demand if quality improvement is implemented), but the option becomes worthless if d is close to zero. Figure 5.3 shows that the option value increases linearly as v (rate of variable production cost savings if a quality improvement program is implemented) increases from 0 to 4 percent. Figure 5.4 shows that the option value decreases with increasing K (implementation cost), and approaches zero when the implementation cost of the quality improvement program is above $100,000 per month.

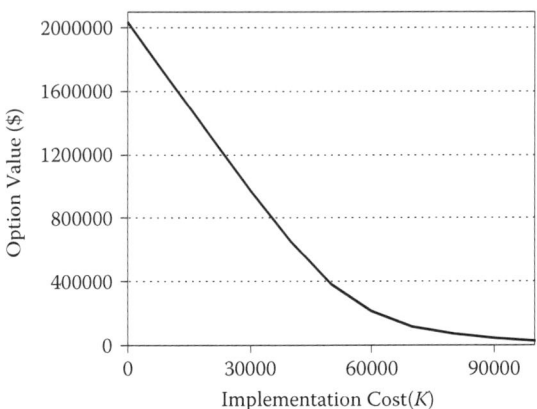

FIGURE 5.4 Option value against K.

5.5 SUMMARY

Quality has become one of the most important consumer decision factors in the selection among competing products and services. There is a substantial return on investment from improved quality, and from successfully employing quality as an integral part of an overall business strategy. By implementing quality improvement programs with statistical tools, companies can improve the quality of their processes, products, and services. As a result, they can reduce their manufacturing or service costs, and they can increase the number of their customers with higher-quality products and brand reputation. More economic benefits of higher-quality products include reduced product recall rates and lower product warranty costs for companies.

Real options valuation is an effective tool to determine the economic benefit potential of quality improvement programs. A real options framework was presented to value a quality improvement program and to determine strategic decisions about the program. A financial model was developed to present the economic benefits of the program. A multinomial lattice technique was used to value the expected benefit potential of the quality improvement program. Sensitivity of the option value against the changes in parameters of the quality improvement program was also investigated.

REFERENCES

Breyfogle, F. W. 2003. *Implementing Six Sigma,* 2nd ed. New York: John Wiley.
Kamrad, B., and P. Ritchken. 1991. Multinomial approximating models for options with k state variables. *Management Science* 37(12): 1640–1653.
Nembhard, H. B., L. Shi, and M. Aktan. 2002. A real options design for quality control charts. *The Engineering Economist* 47(1): 1–32.
Papadakis, E. P. 2007. *Financial justification of nondestructive testing: Cost of quality in manufacturing.* Boca Raton, FL: CRC Press, Taylor & Francis Group.
Persse, J. 2006. *Process improvement essentials.* Sebastopol, CA: O'Reilly Media.

6 Real Options in Outsourcing

Mehmet Aktan
Atatürk University

Harriet Black Nembhard
Pennsylvania State University

Leyuan Shi
University of Wisconsin–Madison

CONTENTS

6.1 INTRODUCTION

Outsourcing offshore for manufacturing firms is the practice of contracting work in other countries to reduce costs, enhance quality, and improve productivity. The advance of technology, the sophistication of business operations, and the need for constant growth are circumstances that suggest a focus on functional core competencies. As companies struggle to adapt to and keep up with the demands of customers and shareholders alike, that focus on core competencies may suggest outsourcing as a potential strategy to remain competitive. The number one reason for outsourcing usually is achieving cost savings, while improving quality, improving time to market, gaining new skills, improving cost predictability, increasing market penetration, and gaining new industry expertise are also among primary drivers for outsourcing (Dominguez 2006).

Outsourcing is made up of two words—"out" and "sourcing." Sourcing refers to the act of transferring work, responsibilities, and decision rights to someone else. Organizations source out work when there are others who can do it cheaper, faster, and better. They source work that can be conducted by others at lower cost and with greater effectiveness or they will waste valuable resources in the pursuit of capabilities that can be readily purchased from others. This pursuit results in poor management since, by its very nature, management is the work of achieving objectives in an effective manner utilizing the least resources. Moreover, engaging in outsourcing allows an organization access to expertise, knowledge, and capabilities found outside its bounds (Power et al. 2006).

Whether or not to outsource is the decision of whether to make or buy. Organizations are continuously faced with the decision of whether to expend resources to create an asset, resource, product, or service internally or to buy it from an external party. An outsourcing initiative calls for the transfer of factors of production and the resources used to perform the work. The organization transferring these is referred to as the client, and the organization that conducts the work is referred to as the vendor (Power et al. 2006).

Manufacturing operations have always been required to adapt to a changing environment. The change presents significant challenges for many aspects of an operation. One strategy has been to increase the flexibility of the manufacturing operation. Flexibility allows firms to compete more effectively in a world of short product life cycles, rapid product development, and substantial demand and/or price uncertainty.

An important issue in applying real options to manufacturing enterprises is to incorporate the practical operational concerns of the decision making into the existing real options framework (Brach 2003). Financial options can be exercised instantly; the trader simply tells the broker to move the contract, and the deal is done. In the existing body of literature on applying real options, the common assumption is that a real option can be implemented immediately as well. However, in complex manufacturing operations, this assumption does not truly hold. Manufacturing operations need some time to be executed. When a decision is made to exercise a real option, there is some period of time until the decision can be implemented (Nembhard et al. 2005).

In this chapter, optimal decision making and value of outsourcing flexibility are investigated considering volatile currency exchange rates between client and vendor countries. Implications of decision implementation time lag on real options valuation are also considered. The strategy is selected among supplying and/or producing at the home country, and outsourcing offshore. The firm maximizes its expected discounted value through the exercise of the manufacturing options in an environment of uncertain exchange rates. The binomial lattice approach is used to model the exchange rate movements and the time lag between the decision and the implementation of the option. Using the procedures presented, a firm can get better estimates for the option value and find the optimal decisions. Without considering the impact of time lag, the value of the operational flexibility can be significantly overestimated.

There has been research in the financial engineering literature that is relevant to this problem. Nembhard et al. (2005) investigated a similar problem where a firm always outsources material and work, and selects the vendor among two foreign countries. Kogut and Kulatilaka (1994) modeled the operating flexibility to shift production between two manufacturing plants located in different countries using a stochastic dynamic programming model. Huchzermeier and Cohen (1996) developed a stochastic dynamic programming formulation for the valuation of global manufacturing strategy options with switching costs, where a firm maximizes its expected, discounted, global, after-tax value through the exercise of product and supply chain network options through the exploitation of flexibility contingent on exchange rate realizations. Dasu and Li (1997) studied the structure of the optimal policies for a firm operating plants in different countries. They determined the structure of the optimal policies for deciding when and by how much to alter the production quantities. Nembhard et al. (2003) studied the option value of being able to switch between the states of producing or outsourcing an item using a Monte Carlo simulation without switching costs between strategies.

The remainder of this chapter is organized as follows. Section 6.2 presents the formulation of the model. Valuation procedure, the solution technique, and numerical results are presented in Section 6.3. Some concluding remarks are in Section 6.4.

6.2 MODEL FORMULATION

Let us consider a supply chain network model for a firm that has a supplier of parts, an assembly plant, and a sales market in a foreign country. The firm can also manufacture the parts and assemble them at home, and sell the finished product at the home country. Parts flow from the suppliers to the assembly plants. They are turned into finished products in the plants, and then the finished products are transferred to the market regions at home and/or the foreign market. The firm defines a set of manufacturing options, where the supplier, plant, and market regions are selected. Each option defines which supplier, plant, and markets of the supply chain network will be chosen. The strategy of the firm is to switch the supplier, plant, and markets according to opportunities presented when the currency exchange rate changes.

The decision maker is concerned with manufacturing flexibility through this strategy as well as knowing the impact of making a switching decision when there is a time lag prior to its implementation. For example, when the firm wants to outsource the parts, the supplier cannot send all of them instantly; when the assembly operation is switched to outsourcing, the required numbers of assembled products cannot be completed instantly in the selected plant; and when a switch is desired for the market regions, the final product cannot reach the new market instantly.

A time horizon is determined for the problem, and it is divided into time intervals of equal length. In each time interval, the firm will select one of the manufacturing options. If there is a switch between the previous and the current decisions, this results in a switching cost. Each switching cost is defined depending on which supplier, plant, and market regions must be changed to exercise the switch.

First, it is assumed that a switch cannot be implemented in the time interval that the decision is given; rather, it will be implemented in the next time interval (i.e., with one time step lag). In the second analysis, it is assumed that a portion of the new decision will be implemented in the time interval that the decision is given, and the remaining portion will be implemented in the next time interval.

Let us formulate a model for the option valuation problem under a time lag as a stochastic dynamic program. This model builds upon the structure of the

Huchzermeier and Cohen (1996) model. The following notation is used for modeling the problem:

t = time index for $t = 0, 1, \ldots, T$

e_t = exchange rate for the foreign country at time t

O_t = option implemented at time t

$P_t(e_t, O_{t-1}, O_t)$ = profit in time interval beginning at time t, if option O_{t-1} was implemented at time $t-1$, option O_t is implemented at time t, and currency exchange rate at time t is e_t

$V_t(e_t, O_{t-1})$ = total expected profit between time t and T, given that the currency exchange rate is e_t, and option implemented at time $t-1$ is O_{t-1}

$W_{O_{t-1}O_t}$ = cost of switching from option O_{t-1} to option O_t

a_i = cost of parts per unit product from supplier i

c_i = unit assembly cost at plant i

R_i = price of the finished product in market region i

x_{spt} = number of units sent from supplier s to plant p in time interval t

x_{pmt} = number of units sent from plant p to market m in time interval t

In each state, the profit is maximized by selecting an option O_t, given that the firm has selected option O_{t-1} in the preceding time interval, and the current currency exchange rate between the home country and foreign country is e_t. The value V_t of the total profit at time t and state (e_t, O_{t-1}) is defined with the following recursive equation:

$$V_t(e_t, O_{t-1}) = \max_{O_t} E[P_t(e_t, O_{t-1}, O_t)] + e^{-r\Delta t}V_{t+1}(e_{t+1}, O_t) \qquad (6.1)$$

In Equation (6.1), O_{t-1} is the option applied in time interval $t-1$, O_t is the option selected for the time interval t, $E[P_t(e_t, O_{t-1}, O_t)]$ is the expected profit during time interval t, and $V_t(e_t, O_{t-1})$ is the total expected profit from time interval t to the last time interval.

Denoting the cost of switching from option O_{t-1} to option O_t as $W_{O_{t-1}O_t}$, and suppressing the time subscript t, we define the profit $P_t(e_t, O_{t-1}, O_t)$ as follows:

$$P_t(e_t, O_{t-1}, O_t) = \sum_p \sum_m (e_m R_m - e_p c_p)x_{pm} - \sum_p \sum_s e_s a_s x_{sp} - W_{O_{t-1}O_t} \qquad (6.2)$$

where subscripts s, p, and m are the set of suppliers, plants, and markets included in the option O_t, respectively.

An advanced form of the time lag problem is the one where a portion of the new option is implemented in the time interval that the decision is given. Let Q be the portion implemented in the current time interval, where $0 \le Q \le 1$. With a lag of one time interval, and with the portion Q of the option implemented in the time interval of the decision, the recursive equation for value V of the total profit at time interval t is defined as follows:

$$V_t(e_t, O_{t-1}) = \max_{O_t} QP_{t-1}(e_{t-1}, O_{t-1}, O_t) + (1-Q)E[P_t(e_t, O_{t-1}, O_t)] + e^{-r\Delta t}V_{t+1}(e_t, O_t) \qquad (6.3)$$

Currency exchange rate between the home country and the foreign country is the source of uncertainty and volatility in the problem. We assume that the exchange rate e_t between the home country's currency and the foreign country's currency follows geometric Brownian motion as

$$\frac{de_t}{e_t} = \mu dt + \sigma dz \qquad (6.4)$$

where μ is the drift of the exchange rate changes, σ is the volatility of the exchange rate, and dz is a standard Wiener disturbance term. The expected changes in the exchange rates are set to $\mu = \exp[(r - r_f)\Delta t]$, where Δt is the length of each time interval, and r and r_f are the risk-free rates of return in the home country and foreign country, respectively.

In the next section, a binomial tree that can be used to model the exchange rate movements is discussed, and the procedure for valuing the problem with implementation time lags is presented.

6.3   VALUING REAL OPTION FOR OUTSOURCING WITH IMPLEMENTATION TIME LAG

The state variable in this problem is the currency exchange rate between the home and the foreign country. In order to model the exchange rate movements and value the options, we can use a binomial lattice technique. If there were no switching costs between the options, each time interval could be evaluated independently from other time intervals. Then, we could simply maximize the profit in each time interval without considering the decisions in other time intervals. Valuing a real options problem with switching costs is more difficult since options exercised in successive time intervals have connections with each other, and the current decision influences the later ones. Hence, we cannot separate the problem into time intervals where the decisions are independent from each other. When switching costs exist, we need to apply a dynamic programming approach to maximize the profit.

Figure 6.1 illustrates the supply chain network with suppliers of parts, assembly plants, and markets in the home and foreign countries. There is one supplier and one plant in each country. At the same time, each country is a market region for the final product. The options are defined based on a number of possible connections in the supply chain network.

Figure 6.2 shows a set of twelve manufacturing options that are defined considering the supply chain network in Figure 6.1. Each option shows the connections among the suppliers, plants, and market regions. For example, the firm may initially be operating under option 7, where the foreign supplier provides parts for the plant in the home country, and the finished products are sold in the foreign market. Later, it may decide to switch to option 9, where the supplier at the home country provides parts for the plant in the home country, and the finished products are sold in both home and foreign markets.

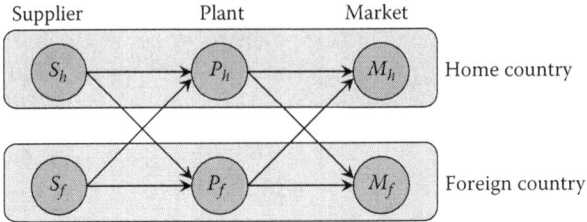

FIGURE 6.1 The supply chain network with two suppliers, plants, and markets.

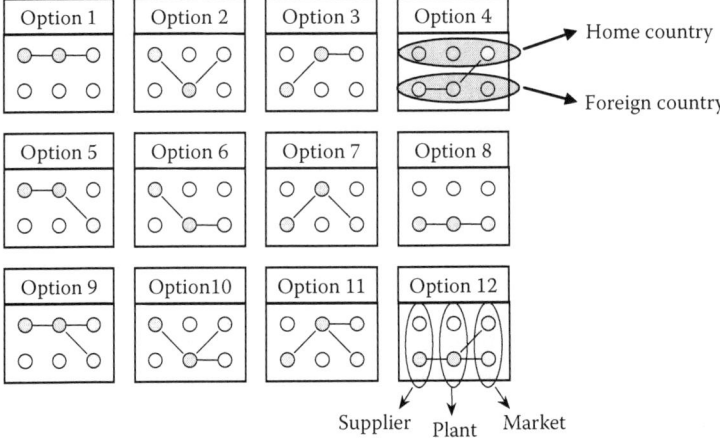

FIGURE 6.2 Twelve manufacturing options for suppliers, plants, and markets.

6.3.1 BINOMIAL LATTICE APPROACH

Assume that the exchange rate e can move up with the rate of u or move down with the rate of d such that $d = 1/u$ at each time interval of length Δt. The rate of movement for the currency exchange rate between the home country and the foreign country is given as follows (Cox et al. 1979):

$$u = e^{\sigma\sqrt{\Delta t}}$$

Figure 6.3 presents the states in the binomial lattice for two steps. There are two states emerging from each state. The element in each node is the state of exchange rate e_t. When we assume that a decision will be implemented in the next time interval, the decision given in a node must be implemented in the immediately following two nodes. At each node, we must select the best option that can be implemented in the next time interval. In order to do that, we select the option that maximizes the recursive function in Equation (6.1) for the expected profit in the immediately following two nodes. Since the value of the state variable in a node is equal to the discounted expected value of the states in the immediately following nodes under the "no arbitrage" principle in option valuation, the profit-maximizing decision for a

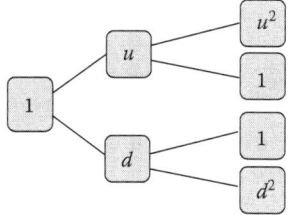

FIGURE 6.3 Currency exchange rates in the lattice.

node is in fact identical to the decision maximizing the expected profit in the immediately following nodes.

For the dynamic programming formulation, we need to select the option that yields the largest expected profit for every possible previous option. We begin from the last time interval and go back one time interval at each iteration. When we reach the first node, the optimal solution can be determined. The complete set of decisions maximizing the total expected profit is obtained by backtracking in the dynamic program.

If there is no time lag to implement the decision, the recursive equation maximizing the profit $V_t(e_t, O_{t-1})$ at time interval t in a lattice is defined as

$$V_t(e_t, O_{t-1}) = \max_{O_t}[P_t(e_t, O_{t-1}, O_t) + e^{-r\Delta t}\sum_{j=1}^{n} p_j V_{t+1}(e_{t+1,j}, O_t)] \tag{6.5}$$

where n is the number of jumps from each node, O_{t-1} is the option in the preceding node at time interval $t-1$, e_t is the exchange rate in the current node at time interval t, and $e_{t+1,j}$ is the exchange rate in node order j of the following n nodes at time interval $t+1$.

With a time lag of one time interval, the recursive equation maximizing the profit $V_t(\bar{e}_t, O_{t-1})$ at time interval t is defined as

$$V_t(\bar{e}_t, O_{t-1}) = \max_{O_t}\sum_{j=1}^{n} p_j[P_t(e_{t,j}, O_{t-1}, O_t) + e^{-r\Delta t}V_{t+1}(\bar{e}_{t+1,j}, O_t)] \tag{6.6}$$

where $\bar{e}_t$ is the set of exchange rates in a group of n nodes at time interval t, $e_{t,j}$ is the exchange rate in node order j of these n nodes at time interval t, and $\bar{e}_{t+1,j}$ is the set of exchange rates in the n nodes that follow the node that has the exchange rate $e_{t,j}$. Hence, option O_t that is selected at time interval $t-1$ will be implemented during time interval t. The selected option O_t maximizes the total expected profit in the n nodes at time interval t and also the following nodes until the last time interval.

With a lag of one time interval, and with a portion Q of the option implemented in the time interval of the decision, the recursive equation maximizing the profit $V_t(\bar{e}_t, O_{t-1})$ at time interval t is defined as

$$V_t(\bar{e}_t, O_{t-1}) = \max_{O_t}QP_{t-1}(e_{t-1}, O_{t-1}, O_t) + \sum_{j=1}^{n} p_j[(1-Q)P_t(e_{t,j}, O_{t-1}, O_t)$$

$$+ e^{-r\Delta t}V_{t+1}(\bar{e}_{t+1,j}, O_t)] \tag{6.7}$$

where portion Q of the option O_t that was selected at time interval $t - 1$ is implemented in time interval $t - 1$, and the remaining portion $(1 - Q)$ of the option O_t is
implemented in time interval t.

Because of the "no arbitrage" principle in the lattice, the discounted expected
value of a state variable for one time period will be equal to the initial value of that
variable, that is,

$$e_t = e^{-r\Delta t} \sum_{j=1}^{n} p_j e_{t+1,j}$$

Therefore, apart from a discount factor of a possible switching cost, the expected
profit with the optimum decision given at time t and implemented with a time lag
at $t + 1$, and the profit with the optimum decision given at time t and implemented
immediately at time t, are in fact equal. Hence the optimum strategy that will be
implemented at time $t + 1$ in the time lag problem, and the resulting profit value, can
both be determined by finding the optimum decision for time t in the original lattice
without time lags. Profit at the first time interval for the time lag problem can be calculated using the given strategy that is implemented at the first time interval.

Assume that the home country is the United States, and the total time horizon of
switching possibilities among the options is three years. Twelve time intervals are
defined, which implies that each time interval is three months. In each time interval,
the firm will exercise one of the twelve manufacturing options given in Figure 6.2.
Assume that the initial parameters are as follows:

$r = 3$ percent (risk-free interest rate in the home country)
$r_f = 4$ percent (risk-free interest rate in the foreign country)
$a_1 = \$1,100$ (cost of parts per unit product at home)
$a_2 = \$1,000$ (initial cost of parts per unit product at foreign supplier)
$c_1 = \$1,050$ (unit assembly cost at home plant)
$c_2 = \$1,000$ (initial unit assembly cost at foreign plant)
$R_1 = \$2,300$ (price of the product in home market)
$R_2 = \$2,200$ (initial price of the product in foreign market)
$D_1 = 1,100$ products/quarter (demand in home market)
$D_2 = 1,000$ products/quarter (demand in foreign market)
$\sigma = 0.3$ (volatility for currency exchange rate)

Switching costs $W_{O_{t-1}O_t}$ between the options are listed in terms of $1,000, as shown
in Table 6.1.

Figure 6.4 shows the expected maximum profits obtained with the lattice for
no-switching and costly switching scenarios. If there is no switching, the same
option will be used at all time intervals. However, when switching is possible, the
best option will be selected at each time interval considering the switching costs.
Then, the value of being able to switch the outsourcing strategy is the difference
between the expected profits of costly switching and no switching. The upper

TABLE 6.1

Switching Costs among the Options

		To											
		1	**2**	**3**	**4**	**5**	**6**	**7**	**8**	**9**	**10**	**11**	**12**
	1	0	120	120	220	100	200	200	300	120	220	220	320
	2	130	0	230	120	210	100	310	200	230	120	330	220
	3	130	230	0	120	210	310	100	200	230	330	120	220
	4	240	130	130	0	320	210	210	100	340	230	230	120
	5	100	200	200	300	0	120	120	220	120	220	220	320
From	**6**	210	100	310	200	130	0	230	120	230	120	330	220
	7	210	310	100	200	130	230	0	130	230	330	120	220
	8	320	210	210	100	240	120	130	0	340	230	230	120
	9	100	200	200	300	100	200	200	300	0	120	120	220
	10	210	100	310	200	210	100	310	200	130	0	230	120
	11	210	310	100	200	210	310	100	200	130	230	0	120
	12	320	210	210	100	320	210	210	100	240	130	130	0

line in Figure 6.4 shows the expected profits when switching between options is possible and the decision can be implemented in the current time interval. The middle line shows the expected profits when switching between options is possible but the decision can be implemented only in the next time interval. The lower line shows the expected profits when there is no switching flexibility. The horizontal axis represents the strategy that was implemented prior to the initial node in the lattice.

If the decisions can be implemented only after a time lag, the value of flexibility will be overestimated when we ignore this limitation. The amount of overestimation is the distance between the upper line and the middle line in Figure 6.4. When a

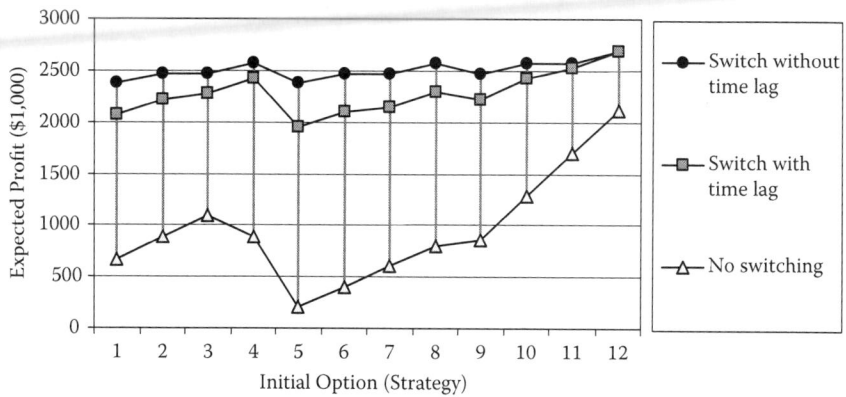

FIGURE 6.4 Expected maximum profits with and without switching flexibility.

decision cannot be implemented in the first time interval immediately, we must implement the strategy that was selected before the first time interval, whereas a strategy can be selected and implemented immediately in the first time interval when there is no time lag. Therefore, if the strategy selected before the first time interval is not the optimal strategy for the first time interval, then there is a profit loss because of implementation time lag. In addition, the best strategy can be selected and implemented at the last time interval when there is no time lag, but a strategy that was selected earlier without observing the last state will be implemented in the last time interval when there is time lag. This is the other source of profit loss when there is implementation time lag.

When there is no implementation time lag for the options, the estimated value of manufacturing flexibility is the distance between the upper and the lower lines in Figure 6.4. When there is implementation time lag, the estimated value of manufacturing flexibility is the distance between the middle and the lower lines in Figure 6.4.

Figure 6.5 shows the expected profits for $Q = 1$, $Q = 0.5$, and $Q = 0$. A value of $Q = 1$ means that the selected strategy can be fully implemented in the current time interval (i.e., there is no time lag between the decision and the implementation), and we get the expected profits shown in the upper line for $Q = 1$. A value of $Q = 0.5$ means that 50 percent of the selected strategy can be implemented in the current time interval, and we get the expected profits shown in the middle line. A value of $Q = 0$ means that the selected strategy cannot be implemented in the current time interval (i.e., the switch can be implemented in the next time interval, so there is a time lag of one time interval) and we get the expected profits shown in the lower line.

With an increasing number of time intervals, the decrease in expected profit due to implementation time lag will become relatively smaller against the total expected profit. Since only four time intervals were evaluated in this analysis, the ratio of expected profit loss due to implementation time lag to the total expected profit is large. The ratio of profit loss to the total profit will get smaller with an increasing number of time intervals, because the difference due to implementation time lag comes only from the first and the last time intervals.

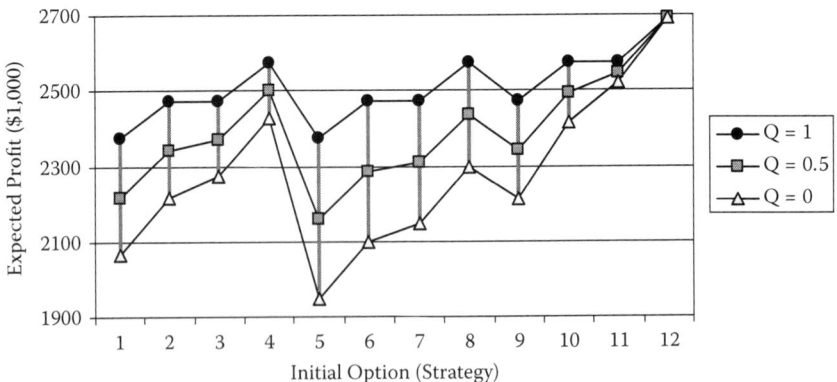

FIGURE 6.5 Expected maximum profits for various strategy implementation time lags.

6.4 SUMMARY

Flexibility allows firms to compete more effectively in an environment of uncertainty. A real options approach estimates the value of flexibility and is used to determine the optimum strategy to manage the flexibility. In this chapter, practical operational concerns of decision making for manufacturing enterprises were incorporated into the existing real options framework by considering the flexibility of outsourcing the parts and the assembly operations, and the currency exchange rate between the home and the vendor country. The difference between immediate implementation of the selected decision and a time lag for the implementation was included in the framework in order to capture and analyze its impact on outcome and the managerial course of action. It was found that when there is time lag for implementation of the selected strategy, the value of flexibility is smaller than in the case of immediate implementation. However, the necessity of analyzing the effect of implementation time lag decreases with an increasing number of decision points in the analysis time horizon.

The real options valuation approach discussed in this chapter gives decision makers a way to choose the appropriate manufacturing enterprise strategy based on an integrated view of the market dynamics. We have presented the application of the procedures on a supply chain network with uncertain exchange rates, where costly switching decisions for the parts supplier, assembly plant, and market regions require time lag to implement. Overall, the manufacturing enterprise maximizes its expected discounted profit through effective decisions to exercise outsourcing options across the supply chain. In such problems, there are multiple options. At each time interval, switching the strategy is possible, and each strategy change results in a switching cost.

REFERENCES

Brach, M. A. 2003. *Real options in practice*. Hoboken, NJ: John Wiley.

Cox, J. C., S. A. Ross, and M. Rubinstein. 1979. An option pricing: A simplified approach. *Journal of Financial Economics* (7): 229–263.

Dasu, S., and L. Li. 1997. Optimal operating policies in the presence of exchange rate variability. *Management Science* 43(5): 705–722.

Dominguez, L. R. 2006. *The manager's step-by-step guide to outsourcing*. New York: McGraw-Hill.

Huchzermeier, A., and M. A. Cohen. 1996. Valuing operational flexibility under exchange rate risk. *Operations Research* 44(1): 100–113.

Kogut, B., and N. Kulatilaka. 1994. Operating flexibility, global manufacturing, and the option value of a multinational network. *Management Science* 40(1): 123–139.

Nembhard, H. B., L. Shi, and M. Aktan. 2003. A real options design for product outsourcing. *The Engineering Economist* 48(3): 199–217.

Nembhard, H. B., L. Shi, and M. Aktan. 2005. A real-options-based analysis for supply chain decisions. *IIE Transactions* (37): 945–956.

Power, M. J., K. C. Desouza, and C. Bonifazi. 2006. *The outsourcing handbook: How to implement a successful outsourcing process*. London: Kogan Page.

7 Barriers to Real Options Adoption and Use in Architecture, Engineering, and Construction Project Management Practice

David N. Ford
Texas A&M University

Michael J. Garvin
Virginia Polytechnic Institute and State University

CONTENTS

The limited adoption and use of real options by practicing managers in the architecture/engineering/construction (AEC) industry remain an important challenge. This chapter describes a risk-rich managerial practice in which real options can add value but are not fully exploited. This setting is used as a basis for identifying and describing specific barriers to widespread real options adoption and use by practicing project managers. These barriers are used to suggest tools, methods, and approaches that may reduce those barriers.

7.1   ARCHITECTURE, ENGINEERING, AND CONSTRUCTION (AEC) ENVIRONMENT

Investments in real assets such as manufacturing facilities or power plants often create future growth opportunities (e.g., follow-on development if demand is favorable) or have contingency possibilities (e.g., delaying or abandoning the project). Similarly, the designers of real assets can improve the ability of systems to respond to future changes through strategies such as modularization. Indeed, these sorts of opportunities or options are prevalent in the built environment. Architecture, engineering, and construction (AEC) projects are rife with uncertainty and ripe with flexibility, which is often incorporated as an intuitive managerial approach for dealing effectively with uncertainty. Examples include preliminary planning and feasibility studies such as environmental impact studies, geotechnical surveys, and traffic volume analyses, which can reveal information that may alter further investment and development decisions. Flexible designs such as oversized foundations and columns permit AEC projects to more readily adapt to changing conditions, such as an increase in expected demand for the project's output. Construction activities often proceed in a series of stages that seek to better define project scope and discover unknown information. Indeed, staged construction, particularly in large projects, can afford decision makers the opportunity to gain more information as market and project conditions become more certain (Miller and Lessard 2000). In short, flexibility can effectively reduce life cycle costs by allowing a timelier and less costly response to a dynamic environment. Undoubtedly, flexibility can add value, but it also comes with costs measured in terms of money, time, and complexity.

Some researchers have identified the potential of real options to improve the management of project uncertainty. Both the amount and nature of project uncertainty make it difficult to plan for and to manage. Miller and Lessard's (2000) study of sixty large ($985 million average cost and 10.7 years average duration) engineering projects concluded that project success depended largely on the amount of uncertainty and how these uncertainties were managed. Ford and Ceylan (2002) investigated the complex nature of uncertainty in a single, large ($2.4 billion) U.S. Department of Energy acquisition project and concluded, in part, that the complexity of managing uncertainty in practice currently exceeds the ability of available tools and methods. Other researchers have extended existing real options pricing models to civil infra-structure contexts (Zhao and Tseng 2003; Zhao et al. 2004; Ng et al. 2002; Ng and Bjornsson 2001; Ho and Liu 2003; Chareonpornpattana et al. 2004; Cui et al. 2004; Moel and Tufano 1999; Ford et al. 2002).

Despite the widespread attention and the general consensus regarding the value of option "thinking," the transfer of real options modeling into AEC project man-agement practice has been slow. Several explanations (described later) have been offered. The writers posit that the attributes of AEC project managers and the char-acter of the AEC industry may have more to do with the lack of transfer of real options analysis than these plausible explanations. Given these circumstances, the identification and assessment of the barriers to real options adoption in AEC project management practice seem to be a prerequisite to devising strategies for the transfer of real options analysis. The subsequent discussion will further characterize AEC project management practice and AEC projects to highlight common characteris-tics and how they reveal barriers to the wider adoption and use of real options. This will lead to the identification and assessment of potential strategies for overcoming these barriers.

7.2   CHALLENGES OF REAL OPTIONS ADOPTION AND USE

As previously mentioned, the transfer of more formal real option modeling into AEC project management practice, as with other fields, has not been rapid. In 2002, a sur-vey of 205 Fortune 1,000 chief finance officers (CFOs) revealed that only 11.4 per-cent use real options, while 96 percent use net present value (NPV; Teach 2003). Several explanations for the slow adoption have been offered. Martha Amram, a well-known authority in real options, has stated, "We've missed something impor-tant. To communicate, [real options analysis] has to be transparent and clear" (quoted in Teach 2003). Alexander Triantis, another leading real options researcher, directly addressed this issue in his five challenges that must be met to take real options from an appealing theoretical concept to a useful practitioner's tool: (1) improve real options models to better reflect reality, (2) understand "split" real options that are owned by multiple agents, (3) model managerial behavior, (4) develop heuristics, and (5) link real option values to the value of the whole firm (Triantis 2005). Each of Triantis's challenges suggests an explanation for the slow adoption of real options by practitioners.* Others (Schmidt 2003; Lander and Pinches 1998; Teach 2003) believe that the cause of slow adoption is a lack of knowledge and understanding of real options by managers and that the education of current and future managers about real options (e.g., in MBA programs) will address this barrier, resulting in more widespread adoption and use. Another rationale for the lack of transfer is the notion that the quantitative models rely upon too many assumptions that may be sensible for pricing financial options but are not reasonable for many real options and therefore cannot or do not price real options accurately. If practicing project managers suspect

---

* See Johnson et al. (2006) for a discussion of how Triantis's five challenges relate to real options in the oil and gas industry.

that this is true, then they are likely to avoid these quantitative models. Yet another rationale concerns the mathematical complexity of some of the models. Many have suggested that the use of sophisticated mathematics such as stochastic calculus limits the accessibility of such models to average decision makers. Without rejecting these potential causes, the current work takes a primarily behavioral perspective of engineering project management practice to identify barriers to real options use. We hypothesize that the attributes of AEC projects, project managers, and the AEC industry may have more to do with the lack of transfer of real options analysis than education or math-based explanations.

Several factors contribute to the challenge of more formal real options use by AEC project managers. First, AEC project managers work in a "risk-rich" environment in which many risks must be managed for project success, many of these risks can cause large decreases in project performance, and risks are interdependent. This situation proliferates practices that focus project management on limiting negative exposure to uncertainty rather than capturing upside potential, and thereby limits the potential benefits of real options use to risk reduction. Second, the management of large-scale AEC projects is quite complex, requiring the management of multiple production activities that routinely employ a wide variety of resources, methods, and technologies; the coordination of labor, materials, and equipment within an environment that is temporary and time constrained; the management of business, environmental, and safety risks before, during, and after projects; and the leadership of a diverse set of stakeholders who often have uncommon interests toward a common goal. A historic example of large-scale project complexity is elegantly documented in McCullough's (1973) account of the development of the Brooklyn Bridge; a more contemporary example is presented by Sabbagh (1991) describing the construction of a modern skyscraper. A structured practice of recognizing, designing, and implementing real options adds significant complexity to project management, so the additional intricacy of real options analysis is not necessarily welcome. Finally, AEC projects are typically unique, and their aforementioned scale often makes them high-cost, long-duration endeavors. Indeed, many projects are multimillion dollar, multiyear enterprises where the opportunity to "get it right" is limited, and little or big mistakes may not necessarily be absorbable for either the AEC manager or organization. In other words, one would be pressed to find an average AEC project. These features and characteristics of the AEC industry make it fertile ground for the use of real options. They also contrast sharply with the financial options and many real options that have been the basis for most real options models. Given these circumstances, the identification and assessment of behavioral barriers to real options adoption in AEC project management practice seem necessary before crafting strategies for the improvement and transfer of real options analysis.

Most of AEC project management of uncertainty focuses on risk. But risk management models (including real options models) include assumptions that do not fully reflect actual project conditions. In the case of real options, these include assumptions of well-behaved future asset values, complete markets for assets, arbitrage opportunities, few and independent options, and the independence of option holders from the future performance of the underlying asset. See Lander and Pinches (1998), Garvin and Cheah (2004), Alessandri et al. (2004), and Ford and Sobek (2005) for descriptions and discussion of real options model inconsistencies with practice. Engineering managers compensate for these gaps by using intuition, judgment, and heuristics to evaluate and select alternatives. Recognizing and understanding the model adaptations used in practice can reveal barriers to the adoption and use of real options. The subsequent discussion will further characterize AEC project management practice and AEC projects to highlight common characteristics and how they reveal barriers to the wider adoption and use of real options. Some of these barriers are challenges that suggest improvements that can increase real options adoption and use. But others appear more like explanations of reasonable, if not perfectly rational in the economic sense, project management practice. The latter are not necessarily problems to be fixed, but conditions to be understood in customizing real options to engineering project management and especially AEC projects.

7.3   PROBLEM CHARACTERISTICS IN AEC PROJECT MANAGEMENT

The structure and culture of the AEC industry create several problems for project management related to the use of real options. We describe three characteristics that generate problems and then address the barriers to adoption that they create.

> *Characteristic 1:* Many AEC project managers have a relatively short-term and local perspective, which tends to promote the conceptualization of risk as exposure to independent events, as opposed to average outcomes across a diverse risk portfolio.

The typical project manager in an AEC organization is overseeing one or a few projects of finite durations at any one time. As used here, an exposure perspective of risk assumes the worst-case scenario will occur and makes decisions to minimize the resulting losses or increase the minimum gains. An exposure perspective of risk contrasts with a probabilistic perspective of risk where decisions are made based upon average (expected) outcomes and managers seek to improve the average. An exposure perspective is (locally) rational because a modest to significant failure in any one project may have very substantial consequences for the project manager's career or professional status (e.g., demotion, dismissal, or loss of professional license). Given this, the natural tendency is to seek to reduce exposure. A bias toward an exposure perspective is exacerbated in situations where the project manager works in an organization that does not have a diversified project and risk portfolio. In such a case, the local perspective is rather appropriate since a significant mistake by a project manager could put the organization in severe financial distress.

Often AEC project managers are rewarded or penalized for performance on individual projects, which reinforces their short-term and local perspective. This is not to say that project managers are not rewarded or penalized based upon cumulative performance but rather both explicit and implicit performance incentives are always present on each and every project. For instance, every time a professional engineer "stamps" a design drawing, that engineer assumes professional liability. A design error can result in additional cost to the engineer (or to his or her firm), a loss of reputation, and/or a loss of professional status, if the error warrants revocation of the engineer's license. Conversely, a superior design can increase the engineer's professional stature, increase professional recognition via industry awards, and/or generate additional income for the engineer (or his or her firm) via new contracts.

> *Characteristic 2:* The centrality and dependence of real options pricing models on uncertainty are inconsistent with the lack of familiarity, understanding, and use of uncertainty by most project managers.

Real options are modeled conceptually and priced formally on the uncertainty of the future behavior, performance, and value of the underlying asset (the project). But project managers do not generally understand uncertainty well or how it can or should impact decision making. Current models cannot turn uncertainty into an effective tool for managing projects instead of a threat to project success. But project management may evolve to make uncertainty an effective management tool. For example, a contractor with an exceptional ability to capture more value of uncertain projects as well as manage construction risks effectively would have a significant competitive advantage. Consider the analogous critical path method for scheduling. For a while after its development in the 1960s, the method's complicated calculations and recalculations of schedule networks and the difficulty of explaining results constrained understanding, acceptance, and use to those with the ability and interest in understanding and performing, checking, and interpreting those calculations. But the development and dispersion of the more intuitive concepts of critical path, total float, and free float and method-based heuristics such as "Feed the critical path" made the method accessible to most project management practitioners, regardless of their interest, understanding, or facility with the calculations. Real options modeling has not yet developed adequately to bridge from the complexity inherent in managing uncertainty with structured flexibility to managerial forms that are useful to a majority of project managers. We later describe how such real options development might occur.

> *Characteristic 3:* The objectives of an AEC project's manager may not necessarily align with the objectives of its owner.

AEC projects usually involve three principal parties: an owner, a designer (architect or engineer), and a builder (general contractor or construction manager). A classic case of conflict is where an architect develops a program for a project that does not quite conform with an owner's intent. This mismatch may be caused by the architect's desire to achieve an artistic effect (which may generate subsequent professional praise and business) that is not shared by the owner or an architect's inability to translate the requirements of an owner into an optimal program. Similarly, a project's builder may prolong construction duration unnecessarily to keep crews busy while awaiting the start of subsequent projects, or the builder may substitute inferior materials to those specified to reduce his or her costs and thereby increase profits. Differences in objectives can cause project managers to not select real options that maximize project value.

AEC projects are inherently unique and complex—combining technical, logistical, financial, organizational, political, and social issues in a temporary engagement among a project's stakeholders. Not surprisingly, this often puts the managers of these projects in very challenging situations. For instance, a routine day during the construction period of a bridge renovation project might involve lane closure and traffic management operations; just-in-time delivery of materials or equipment; multiple concurrent production activities such as decking demolition and removal, replacement of supporting steel members, and installation of new decking; procurement of materials needed in the future; inspection of completed work by a third party; and responding to the project's owner about complaints received from motorists who use the bridge. For a project manager, the challenges of managing complex projects may make simplifying the management effort a priority or at least of equal importance to adding project value. These different objectives can create the challenges of multiple option ownership described in Triantis's second challenge (2005).

7.4 BARRIERS TO REAL OPTIONS ADOPTION AND USE

7.4.1 BARRIER 1: REAL OPTIONS MODELS ASSUME MANY REPEATED BETS, BUT PROJECT MANAGERS MAKE ONE-SHOT CHOICES

Not all options that are modeled, priced accurately and whose price recommends its purchase and use increase project value. Whether an option actually increases project value or not depends on uncertainty resolution and managerial decisions (the application of the exercise decision rule). If the uncertainty resolves such that the option should not be exercised, then the option purchase and maintenance costs are paid without capturing benefits. (Cases in which there is no possibility of uncertainty resolving such that the option should not be exercised are not options since the decision to exercise should be taken immediately.) These options *decrease* project value. These options are recommended because they are priced based on the average payoff of many repeated bets, that is, on the assumption that the same circumstances and option will occur many times and the option holder will capture the average of all the benefits and losses. Engineering project management practice often differs markedly from this assumption. Project managers face many one-shot choices where they will likely experience the circumstances and option only once. This encourages an exposure-based perspective of risk and can create problems for managers of single options.

Consider the simplified example of a construction project that has fallen behind schedule. The expected completion date is 100 days after the original deadline, with a range of possible delays from 50 to 150 days. An option to improve productivity has been obtained that will change the expected completion delay to an average of 75 days with a range of 40 to 125 days. Alternatively, overtime can be used to change the expected completion delay to an average of 85 days with a range of 50 to 100 days. A project manager with an exposure-based perspective of risk (e.g., who fears being dismissed if the project is over 100 days late) may apply a one-shot perspective and use the single worst possible conditions to select a strategy instead of a probabilistic perspective that uses average values. In this example, this would result in the manager using overtime instead of improved productivity because overtime reduces project exposure more than improved productivity, from 150 days to 100 days instead of from 150 days to 125 days, even though improved productivity would improve the average project schedule performance more than overtime, from 100 days to 75 days instead of from 100 days to 85 days. Ceteris paribus (all other things held equal), the productivity option is priced higher than the overtime strategy. But the manager rationally chooses overtime based on his or her one-shot, exposure-based perspective that precludes waiting for the average payoff.

Research supports the existence and common use of an exposure-based perspective of risk. The results of controlled experiments by Li (2003) support our distinction between exposure-based and probabilistic perspectives in decision making under uncertainty. In these experiments, subjects preferred a chance with a lower expected payoff and a higher minimum payoff (i.e., less exposure) to a chance with a higher expected payoff and a larger chance of no payoff. The subjects took an exposure-based perspective by preferring to improve their worst-case scenario (minimize their loss) instead of maximizing their expected value even though they were provided and understood the reward structure. Li (2003, p. 122) concludes, in part, "These results therefore raise doubts about whether people behave . . . as if they were always trying to apply a long-run [probabilistic] perspective of maximization to short-run [one-shot] probabilistic events."

What might cause a project manager to take either an exposure-based or probabilistic perspective and make the very different decisions that each suggests? First, project managers may intuitively foresee circumstances in which they might be forced to explain an expense they authorized (to obtain or maintain an option) that, in hindsight, clearly did not add project value because the option was not needed. Using expected values allows valuation with uncertain futures but also makes the value added by any one specific option uncertain. Project managers may avoid strategies that are difficult to defend, even if they can, but may not, add value. Second, assuming that the manager accurately prices the options, the choice of an exposure or probabilistic perspective may depend upon whether they can survive the potential losses that may occur if either the option is not exercised or wrongly exercised, or if losses exceed benefits before the law of averages evens out and the long term net value of options is realized. This choice depends largely upon the manager's risk tolerance and the incentive structures used by an enterprise for its project managers. If the common short-term or project-specific incentives are dominant, the project manager is likely to act conservatively and only execute low-risk, low-payoff options since overall performance and job security are strongly linked with performance on independent and mutually exclusive projects. In contrast, if longer-term and cumulative project incentives are dominant, then project managers can probably rely more on the law of averages in decision making. Hence, in this circumstance more high-risk, high-payoff options are likely to be implemented.

One argument against an exposure perspective being a rational barrier to real options adoption is that the option should be considered similar to insurance, in that the cost (premium) is justified for the protection (coverage) whether or not the option is exercised (claim made) or not. This is essentially an argument for a probabilistic perspective. However, to remain competitive in competitive bidding circumstances, firms often cannot afford to include the cost of loss-limiting (i.e., put) options and remain competitive because often a competitor will assume the uncertainty will resolve in the desired way such that the high expenses will not occur or that they can be recovered, such as through a deadline extension. Such a "hope-for-the-best" practice of not including adequate protection for uncertain conditions is fatal to overly optimistic firms in the long run (over many projects). But the potential to shift costs to others (e.g., through change orders due to unexpected conditions) and low barriers to entry for some parts of the industry maintain a population of such competitors, preventing more cautious firms from including reasonable put options.

7.4.2   BARRIER 2: PROJECT MANAGERS ARE RISK AVERSE IN VALUING REAL OPTIONS

Like almost all managers, engineering project managers tend to be risk averse, meaning that they are willing to forgo some benefits to reduce uncertainty. Given two otherwise equal strategies, they prefer the one that depends less on the resolution of an uncertainty to determine whether, or how much, it adds value to their project. Many managerial actions in which uncertainty is perceived to make little or no difference in whether value is added or not (e.g., budget increases, or scope decreases) can and do increase project value. In sharp contrast, all real options are, and if perceived accurately are understood to be, very dependent on how uncertain conditions resolve. Practicing managers demonstrate their risk aversion regularly when choosing among value-adding alternatives with different amounts of uncertainty and reward. As an example of managerial risk aversion in practice, a risk-averse manager might prefer to extend a project's deadline to reduce a project's forecasted completion delay instead of adopting a new, untested technology to accelerate production even if the resulting expected delay with the deadline strategy is larger. This could be because the duration reductions of the new technology are considered more uncertain and more likely to fail than the deadline change.

Managers may tacitly implement risk aversion by adding a cost in their valuation of real options that reflects their level of risk aversion. This would decrease the attractiveness of real options relative to more certain alternatives and thereby decreases the use of real options. The lower value that a manager is willing to accept to get certainty is the certain equivalent value of the uncertain strategy (Howard and Matheson 1972). In real options assessment and selection, this difference effectively adds a risk aversion cost to real options that can be as large as the difference between the price of the option and the manager's certain equivalent of that alternative. Current real options pricing models do not include the risk aversion costs due to manager's perspectives. These managerial risk aversion costs increase the perceived cost of real options, reduce their values to managers and therefore real options are used less than their price implies. Discounting real options due to aversion to uncertainty may reflect a manager's perspective of risk as exposure and attempts to reduce exposure.

Risk aversion and the resulting reduced use of real options may be very rational from the manager's perspective. Why would a manager take the risk of being wrong (not needing to exercise the option to add value and thereby reducing project value by the purchase and maintenance costs of the option) when less risk is taken with a more certain alternative? If an exposure-based perspective is appropriate, managerial inclusion of risk aversion costs in real options valuation can be also appropriate. However, the size of those costs is probably determined intuitively and tacitly, may be larger than needed to reflect the value of reduced uncertainty, and may reduce real options values more than they should, thereby reducing real options use.

7.4.3 BARRIER 3: MANAGERIAL PROJECT REAL OPTIONS ARE WORTH LESS THAN TRADITIONAL PRICING SUGGESTS

Project management practice varies greatly from two fundamental assumptions of most real options pricing models. These differences cause real options to be valued less than estimated prices. First, there are many possible managerial actions that increase project value. Examples abound, including overtime or special equipment to control schedule performance, subcontracting work to access additional resources, and shifting targets. Many of these can be structured as options. Trigeorgis (1993) shows with closed form valuations and Bhargav and Ford (2006) demonstrate with simulation that the values of individual options decrease with the number of additional options that are available. One intuitive interpretation of this result is that each additional alternative strategy that improves performance without the option being considered reduces the amount of benefit the option can add. Therefore, the many alternatives for managing uncertainty that are available to project managers may reduce the value of each real option to significantly less than if it was the only alternative. If project managers intuitively understand this relationship and include it in their tacit valuations of real options, they may estimate real options values closer to their actual values than traditional, single-option pricing models. Both the lower estimated values and the discrepancy between formal valuations and perceived values could reduce the acceptance and use of real options by project managers.

A second reason that some project management real options are worth less than traditional pricing suggests is that most option models assume that the option holder does not influence the value of the underlying asset. This assumption is usually unstated in the literature. Its foundation is in option pricing models for financial assets (e.g., stocks in a market that can reasonably be assumed to be perfect) in which the option holder is independent of the asset except through the market. For some real options, this assumption is reasonable. For example, the holder of an option to accelerate exploitation of a fossil fuel reservoir by drilling additional wells cannot influence the characteristics of the reservoir (e.g., size, or porosity) or the market price of the refined products. In sharp contrast, when project managers use real options to control their own projects, they purposefully and strongly contradict the assumption of option holder/uncertainty independence by working to manipulate the uncertainties in their projects through traditional means. Examples of these uncertainty manipulations in project management are numerous, including using overtime or special equipment to control schedule performance, taking subcontracted work in-house, and using construction manager at-risk contracts that include options to change builders. Miller and Lessard (2000) describe these dependencies in major project decisions, and Alessandri et al. (2004) describe this type of linkage in a specific set of project management decisions. In these cases, real options decisions and project management decisions are tightly linked. Therefore, real options pricing models that assume independence of option holders and underlying assets and uncertainties may not price strategies accurately enough to guide AEC project managers. Since project managers tend to manipulate project uncertainties to increase project values, this reduces the potential benefits of options. Therefore, violating this pricing assumption may cause real options to be overvalued using traditional models. If we assume that managers intuitively value options and include these two features of project management practice, this would reduce the value of options, requiring that they add more value to be justified and reducing the use of real options by project managers.

7.4.4   BARRIER 4: PROJECT MANAGERS HAVE LIMITED RESOURCES

Real options theory says that when an option adds value, the potential holder of the option should purchase, maintain, and use the option. But practicing managers often require that options add lots of value before they are purchased. We have heard managers describe the circumstances that justified the purchase of an option as "no brainers" (i.e., the option would probably add so much value that the manager considered the choice to purchase the option to be obvious). Why do practicing managers require very large expected payoffs and regularly forgo obtaining and using options that potentially add value? Reduced actual and perceived option values for the reasons described above can provide a partial explanation. Resource limitations may also play a part.

As described, engineering project managers often have a plethora of alternatives for adding project value. Most of them require resources to identify, design, analyze, and implement. Limitations on several types of resources restrict the use of real options, including (1) funds for purchasing and retaining flexibility; (2) labor, equipment, and materials to implement management decisions; (3) combinations of cognitive ability, tools, and methods to understand, design, evaluate, and implement options and other value-adding alternatives; and (4) time and attention to recognize and use options and other value-adding alternatives. Engineering project managers are forced to choose from among their many alternatives for increasing project value when faced with these constraints. Choices are often based on a benefit–cost ratio analysis to maximize total project value derived from any given set of limited resources. Alternatives with the largest perceived ratio are chosen first. If conditions such as holding many project improvement alternatives (Barrier 3), attitudes such as uncertainty aversion (Barrier 2), and managerial perceptions such as not understanding and therefore avoiding options drive managers' evaluations, then real options will have relatively low benefit–cost ratios and will therefore be selected rarely.

7.4.5   BARRIER 5: AS OPTION HOLDERS, PROJECT MANAGERS DO NOT NECESSARILY SEEK TO MAXIMIZE PROJECT VALUE

Option holders not seeking to increase asset values contrast directly with real options theory and are related to the classic agency problem discussed by organizational behavior theorists. Managerial real options are often valued and assessed in dimensions that cannot be measured with money, or at least with project money. For example, Ford and Ceylan (2002) found that managers at the U.S. Department of Energy's National Ignition Facility explicitly used options to increase the likelihood of successful technology development and to increase vendor competition, as well as to reduce costs. The previously described circumstance where failure seriously harms the manager's career is another good example. Providing protection against project failure and therefore career damage increases the value of the option to the manager, if not the project. Hence, the manager assesses the option differently than the project owner (or its investors) would. Again, the U.S. Department of Energy's National Ignition Facility project provides an example. Here, the project manager linked an unfortunate resolution of uncertainty without a specific option (unsuccessful technology development) with project failure (Ford and Ceylan 2002). The high-profile nature of the project and failures of previous managers made the potential of project failure and its impacts on his career unacceptable to the manager. Therefore, the manager might have valued the option more highly than the project owner (U.S. Congress, in this case), who might have been willing to accept project failure, although there is no direct evidence that this assessment actually took place. This pushes real options adoption and use away from traditional "optimal" and "project-maximizing" choices.

The presence of factors other than economic product value changes the environment for real options analysis substantially. As opposed to the presumed economic or project-centric view, evaluations regarding options are made based upon dimensions other than project money. This can lead to an increase in the perceived value of certain options to the project manager that may not necessarily enhance the value of the project. This also suggests that project managers have various motivations as well as means to manipulate or influence asset values. This may reduce the perceived value of project-centric options.

7.4.6   BARRIER 6: EXERCISING OPTIONS CAN HAVE DRAMATIC SECONDARY IMPACTS ON PROJECT MANAGEMENT THAT INCREASE THE DIFFICULTY OF PROJECT MANAGEMENT

A special but particularly widespread case of limited resources that all project managers and project management teams experience is bounded rationality (a well-known notion first characterized by Nobel Prize winning economist Herbert Simon), their maximum cognitive capacity. Project management tools can expand the capacity, but there remains an upper limit. The complexity inherent in AEC projects often approaches or exceeds the bounded rationality of project managers and project management teams. Therefore (ceteris paribus) project managers prefer simpler alternatives to more complex ones. An alternative with many or complex side effects is less attractive. This can decrease the attractiveness of options that add management complexity from the perspective of the project manager.

Consider the example of an actual situation disguised as the Project Isolated case and described by Johnson et al. (2006). Project Isolated is a new fossil fuel development project in a remote location requiring a specialized piece of equipment that is available from only one manufacturer. Once equipment manufacturing is complete, it will be transported by sealift from the manufacturer to the project location. However, the Project Isolated site is accessible by sea only during a short time window due to weather. The Project Isolated team is concerned that the manufacturer will not complete the equipment in time for delivery to the site by sealift within the available time window. If this window is missed, the next available window is several months later. This would significantly delay the development of Project Isolated and the delivery of product, and therefore severely degrade project performance. The project team is considering purchasing an option to transport the equipment by a more expensive airlift to avoid missing the weather window.

Johnson et al. (2006) developed a relatively simple project simulation model. The model was used to estimate option value. But the option's monetary value reflects only part of the impact the airlift option has on the project and its management. Johnson et al. (2006) also used the model to simulate delivery dates with and without the air-transit option (Figure 7.1).

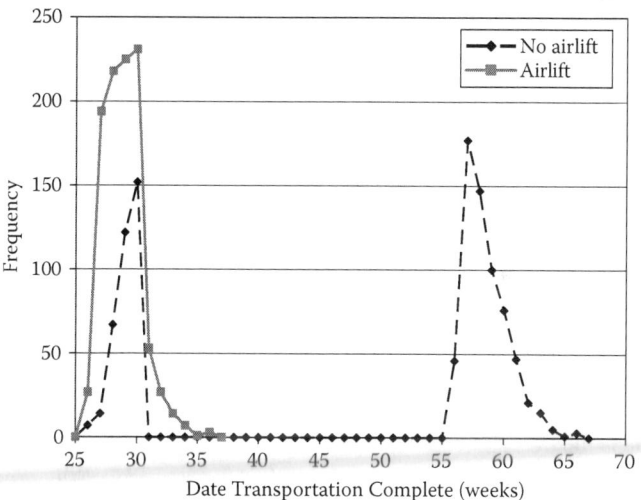

FIGURE 7.1 Distribution of transportation completion dates.

The airlift option has at least two significant impacts on project management. First, the option transforms the transportation completion date distribution from a bimodal distribution into a single modal distribution. This requires that the project be prepared for delivery in one, not two, discrete time periods. Second, without the airlift option the project manager must design and prepare for only one mode of delivery (sealift). But with the airlift option, the project manager must design and prepare for two possible delivery modes (sealift and airlift). Preparing for both sealift and airlift delivery is clearly a more difficult project management task than planning for either one. The combined impacts of the option may increase project management complexity. Faced with potentially exceeding the project management's bounded rationality if an option is used, some managers may find options less attractive and tacitly discount some option values to account for the anticipated additional managerial complexity.

7.5   OVERCOMING BARRIERS TO REAL OPTIONS ADOPTION AND USE

Significant improvement can be made to the adoption and use of real options by AEC project managers by developing and providing tools and methods that expand the capabilities of project managers to understand and use options. Useful tools and methods can be developed from both sides of the understanding and application gap by (1) improving existing real options models to better reflect AEC projects and project management practice, and (2) using existing project management concepts, tools, and methods to model real options. We describe and discuss each of these approaches.

7.5.1   IMPROVE EXISTING REAL OPTIONS PRICING MODELS TO BETTER REFLECT AEC PROJECTS AND PROJECT MANAGEMENT PRACTICE

This approach seeks to improve the accuracy and applicability of real options models for AEC projects (Triantis's third challenge, 2005). The many methods for modeling real options and their applications across different settings have created a confusing situation for project managers, and the confusion is heightened by different assumptions that underlie the modeling techniques. Assumptions are often taken for granted, misunderstood, or unrealistic. Critics are quick to jump to the conclusion that the entire real options field is mathematically elegant, but hardly useful in practice. We disagree, but believe the current state of the art of most real options models remains distant from most AEC project management practice. Several common features of existing real options models differ from current practice to such an extent that improving the models might lead to greater acceptance of real options by practitioners.

One issue is the prevalence of single-factor models—models that use one variable to represent the underlying value of the asset or project over time. Copeland and Tufano (2004) have suggested that very often the underlying value of a real asset is driven by one key variable. Their assertion is probably correct for some relatively simple types of assets such as commodities or natural resources. In these cases, a single-factor model is sufficient to price project options. Pricing techniques for single-factor models are generally familiar (to ROA analysts at least) and relatively simple, such as the binomial tree method. Conversely, many financial analysts of complex infrastructure projects are likely to suggest that it is difficult to describe the drivers of an infrastructure project's value with a single variable. Valuing these complex assets and options to increase their efficacy typically requires multiple interdependent variables. Multifactor models generally require Monte Carlo simulation techniques or more sophisticated lattice models. Until recently, the use of Monte Carlo simulation was possible on only the simplest forms of options (e.g., European); however, recent advances have made its use more practical for other forms of options (Broadie and Glasserman 1997; Longstaff and Schwartz 2001).

For example, Chiara and Garvin (2007) recently introduced a novel approach that models revenue risk mitigation contracts in privately financed infrastructure projects as multiple-exercise real options. The approach structured revenue guarantees (put options) into risk mitigation contracts and valued them by two methods, the multi-least squares Monte Carlo method and the multi-exercise boundary method. Effectively, the two methods combine Monte Carlo simulation and dynamic programming techniques to price the multi-exercise options. While this approach may be more palatable with regard to the evolution of a project's value, the method is clearly more mathematically sophisticated than most traditional real options pricing models. Moreover, it is currently applicable to a very specific type of project option. Thus, it may be more realistic, but this potential improvement comes at a cost in terms of modeling complexity and specificity.

7.5.2   USE EXISTING PROJECT MANAGEMENT CONCEPTS, TOOLS, AND METHODS TO MODEL REAL OPTIONS

The objective of this approach is different from the first. This approach seeks to improve the options thinking skills of practitioners by (1) improving their understanding of and intuition about real options, and (2) providing practical heuristics that reflect both real options theory and practice (Triantis's fourth challenge, 2005). Optimally these models begin with practitioner understanding as it exists and include aspects of projects and project management that interact with real options use. Many options thinking skills can be improved by building upon basic project management concepts, tools, and methods. Examples include (1) recognizing management challenges as uncertainty management situations, (2) explicitly considering flexible project management alternatives that can be structured as options in decision making, (3) assessing when flexibility can and cannot add value, (4) recognizing options in common and innovative project management alternatives and structuring those alternatives as options, (5) developing order-of-magnitude estimating techniques of option values, and (6) creating guidelines for implementing adopted options to capture their value.

A process model of designing real options is a simple example of a tool that can improve options thinking skills. Options development and use can be modeled as a series of steps that include describing the uncertainty to manage, developing management alternatives, structuring alternatives as options, building alternative designs, strategy selection, and option implementation (Mun 2005). A more detailed process model could provide additional guidance and a framework for other tools that improve options thinking skills. One such tool developed by Johnson et al. (2006) for structuring some project management alternatives as options can help project managers to more fully grasp the potential utility of a flexible strategy when compared with an inflexible one. Figure 7.2 depicts the tool, and the subsequent example illustrates its use.

For a given project challenge or situation, to structure a flexible project management alternative as a real option:

1. Write a one- or two-paragraph description of the situation that accurately describes the challenge facing the project team and the available solutions, including the solution that includes flexibility. This description should

include a statement of the challenge, identification of what influences the challenge, a way to measure this influence, and solution strategies for addressing the challenge.

2. Based on the description, complete the following paragraph.

The challenge facing this project is _____(1)_____. The uncertainty that is causing this challenge is _____(2)_____. The traditional approach to this challenge is _____(3)_____. A possible alternate solution to this problem is _____(4)_____. The performance measurement that can be used to evaluate these alternative strategies is _____(5)_____. The value of this measurement that justifies switching from the traditional strategy to the alternative strategy is _____(6)_____. In order to have the ability to change strategies, we must _____(7)_____. To change strategies, we would _____(8)_____.

3. Is the paragraph consistent with the description of the situation? Does the paragraph accurately describe the situation? If not, adjust the paragraph to describe the situation.

4. Fill in the structured description in Figure 7.2 by placing the answers from the paragraph above in the appropriate box. Complete the decision rule as shown.

Uncertain performance measure	1
Driver of performance uncertainty	2
Reference strategy	3
Alternative strategy	4
Signal for changing strategy	5
Conditions for strategy change	6
Actions required to obtain or retain flexibility	7
Action required to change strategy	8
Decision rule for changing strategy	IF (box 5) MEETS (box 6) THEN (box 4) ELSE (box 3)

FIGURE 7.2 A tool for structuring flexible project management strategies as options.

The tool transforms a poorly structured management problem into a real options description. Specifically, (1) is a measure of asset value, (2) is the driving uncertainty, (3) is the strategy without the option, (4) is the strategy if the option is exercised, (5) is the condition that is monitored for making the exercise decision, (6) is the strike or exercise conditions or price, (7) are the actions that generate option purchase and holding costs, (8) are the actions that generate option exercise costs, and the decision rule defines if and when the option should be implemented. The Project Isolated example described in Section 7.4.6 can be structured using the tool described below and shown in Figure 7.3.

The challenge that this project is facing is a possible delay in the start of production. The uncertainty that is causing this challenge is the delivery of equipment to the Project Isolated site. The traditional approach to this challenge is to use a sealift to deliver the equipment. A possible alternate solution to this problem is to airlift the equipment to the Project Isolated site. The performance measurement that can be used to evaluate the strategies is the forecasted delivery date of the equipment by sealift. The value of this measurement that justifies switching from the traditional strategy to the alternative strategy is the required equipment delivery date (close of the weather window). In order to have the ability to change strategies, the Project Isolated team must reserve airlift capacity and design the equipment so that it can be airlifted. To change strategies, the Project Isolated team would cancel the sealift and notify the airlift company.

Uncertain performance measure	Start of production.
Driver of performance uncertainty	Equipment delivery date.
Reference strategy	Sealift equipment.
Alternative strategy	Airlift equipment.
Signal for changing strategy	Forecasted delivery date of the equipment by sealift.
Conditions for strategy change	End of weather window.
Actions required to obtain or retain flexibility	Reserve airlift capacity in advance, design equipment for airlift.
Action required to change strategy	Cancel sealift, notify airlift company.
Decision rule for changing strategy	IF (forecasted delivery date) > (end of weather window), THEN (airlift equipment), ELSE (sealift equipment).

FIGURE 7.3 The structure of an option to airlift equipment.

A second example of a tool that can improve options thinking skills addresses valuation. Practicing managers often use simple valuation tools such as net benefit, benefit-cost ratios, and payback periods. A potentially common error in options thinking is the failure to recognize and include all the relevant benefits and costs of the option. The following "algebra of options" compares option benefits and costs to estimate the net economic value of the option:

$$V = B - C \tag{7.1}$$

where

V: option value or the value added by flexibility
B: benefits provided by the real option
C: costs of the real option

Expanding the benefits and costs into their respective components gives the following:

$$V = (B_p + B_c + B_s) - (B_1 + C_i + C_m + C_e + C_s) \qquad (7.2)$$

where

B_p: net performance benefits, the value of improved performance if the decision is delayed. This value can be estimated as the difference between the project benefits with and without the option.

B_c: cost benefits, the value of reduced costs if the decision is delayed. This value can be estimated as the difference between the project costs with and without the option.

B_s: strategic benefits, the value of delaying decisions that are external to the project.

B_1: lost benefits, benefits that are unavailable due to delaying the decision. This value can be estimated as the sum of the lost performance benefits and the lost cost benefits.

C_i: initial option costs, the costs of initially obtaining the flexibility.

C_m: option maintenance costs, costs of keeping the flexibility available until the delayed decision is made.

C_e: exercise costs, costs to implement a change in the project strategy.

C_s: strategic costs, costs incurred by delaying the decision that are external to the project.

For rigorous analysis, all valuations must be consistent for aggregation, such as having the same time basis (e.g., annual or over the entire project) and being discounted or projected to the same point in time with discount rates that reflect the uncertainty in the discounted or projected values. Other tools such as Feinstein and Lander's weighted average discount rate (Feinstein and Lander 2002) can facilitate this modeling. However, even without completely rigorous analysis, Equation (7.2) helps identify, describe, and evaluate the potential of options to increase project value by providing a disaggregation of option value into components that are relatively easily recognized, described, and discussed by practicing managers. Some costs or benefits may be zero. But missing one could cause a manager to make a worse decision about purchasing an option than if all the benefits and costs were incorporated into the decision. The algebra of options provides a checklist of potential sources of impacts on real options value as well as a specification of the directions of impacts. Likewise, some components may be easier to quantify than others. In some cases, such as when one or two components of the option value far exceed all others, rough estimates may be adequate to make accurate decisions about project management options.

The algebra of options can also help managers decide when to consider an option. Using maximum estimated benefits and minimum costs and vice versa in Equation (7.2) can be useful in "triaging" risk management strategies into three categories: (1) those for which the minimum benefits and maximum costs provide positive net value and therefore should be committed to immediately, (2) those for which the polarity of the net value depends on the benefits and costs and therefore consideration of an option is advised, and (3) those for which the maximum benefits and minimum costs provide negative net value and therefore should not be considered further. The further development of tools such as the algebra of options will facilitate the development of heuristics that managers can use to identify and assess options.

7.6   DISCUSSION

The barriers to the widespread adoption and use of real options by practicing AEC project managers are rooted in the complexity of AEC projects, the limitations and characteristics of managers, and the current state of the art of real options models. These root causes interact to create the barriers, such as the cognitive limitations of managers interacting with the complexity of projects, to create a preference for simpler project management alternatives. The multiplicity and interdependence of the causes of the barriers prevent any one single approach from successfully bridging those barriers. Advancements in multiple areas that borrow from and link different perspectives are needed. We have suggested two perspectives that may prove useful: the improvement of existing real options pricing models toward practice and the use of existing project management concepts, tools, and methods to improve options thinking skills. Each approach can help address at least one barrier, but each approach has limitations.

The improvement of existing real options pricing models to better reflect AEC project management practice (the first suggested approach) can help address the gap between real options models and practice. More specifically, we suspect that the valuation of combinations of different types of real options that have a variety of characteristics (Barrier 3, Section 7.4.3) can be significantly improved through additional real options model development. These models could provide insights and potentially valuable project management heuristics about the use of real options. However, more elaborate valuation models worsen other barriers by adding model complexity and thereby making it harder for practicing managers to understand the model and accept model results. This approach to bridging the barriers sets up a trade-off between model accuracy/applicability and simplicity that facilitates understanding and acceptance. The choice should typically depend upon the purpose of the valuation; if a tactical or reasonably precise analysis is necessary the former approach is probably required, but if a strategic or order of magnitude analysis is needed the latter may be adequate. More than likely, the field of ROA will produce additional and more specialized approaches in the future. Regrettably, the fundamental trade-off between complexity and precision versus simplicity and inaccuracy will probably remain. Although we believe that real options pricing model developments should be pursued, improving those models cannot resolve the real options adoption and use challenge alone.

The second suggested approach to bridging the barriers builds on existing project management concepts, tools, and methods to improve options thinking skills. This approach more directly addresses the primarily behavioral barriers (Barriers 1, 2, 4,

5, and 6). Generally, these barriers result from human motivations and limitations, the industry's project environment, and common organizational policies and norms. To be effective, tools that improve utilization of managerial options may need to be "thinking" tools, not "solution" tools. As used here, thinking tools initiate, facilitate, or guide a manager's cognitive processes and add insight without necessarily providing exact or always correct solutions. The two tools described previously are examples of thinking tools. Consultants already may have developed and provided such tools for improving real options thinking to their clients. But, to our knowledge, few, if any, have been rigorously tested and they are not publicly or widely available, thereby severely limiting their usefulness for bridging the barriers on a large scale. More sharing of existing tools and the development and testing of additional tools and methods for improving options thinking skills are needed.

Fundamental features and characteristics of AEC projects, their managers, and existing real options models have created large barriers to the widespread adoption and use of real options by practicing managers. These barriers prevent or severely limit AEC project managers from capturing the potential benefits of real options. The barriers can be overcome only by broadening the development of real options tools and methods to include, and therefore balance pricing, valuation, project characteristics, and managerial practice. Doing so may be difficult but can transform real options into a standard part of every project manager's toolkit.

ACKNOWLEDGMENT

The authors thank Professor Ken Reinschmidt for his contributions to the concepts of exposure versus probabilistic perspectives of risk.

REFERENCES

Alessandri, T., Ford, D., Lander, D. Leggio, K., and Taylor, M., 2004. Managing risk and uncertainty in complex capital projects. *Quarterly Review of Economics and Finance* 44(5): 251–267.

Bhargav, S., and Ford, D. 2006. Project management quality and the value of flexible strategies. *Engineering, Construction, and Architectural Management* 13(3): 275–289.

Broadie, M., and Glasserman, P. 1997. Monte Carlo methods for pricing high-dimensional American options: An overview. Working paper. New York: Columbia University.

Chareonpornpattana, S., Minato, T., and Nakahama, S. 2004. Government supports as real options in built-operate-transfer projects. http://www.realoptions.org/papers2003/CharoenMinatoNakahama.pdf.

Chiara, N., and Garvin, M. J. 2007. Utilizing real options for revenue risk mitigation in transportation project financing. Transportation Research Record No. 1993. Washington, DC: Transportation Research Board of the National Academies.

Copeland, T., and Tufano, P. 2004. A real-world way to manage real options. *Harvard Business Review* 84(3): 90–99.

Cui, Q., Bayraktar, M., Hastak, M., and Minkarah, I. 2004. Uses of warranties on highway projects: A real option perspective. *Journal of Management in Engineering* 20(3): 118–125.

Feinstein, S. P. and Lander, D. M. 2002. A Better Understanding of Why NPV Undervalues Managerial Flexibility. *Engineering Economist* 47(4): 418–435.

Ford, D. N., and Ceylan, K. 2002. Using options to manage dynamic uncertainty in acquisition projects. *Acquisition Review Quarterly* 9(4): 243–258.

Ford, D. N., Lander, D., and Voyer, J., 2002. The real options approach to valuing strategic flexibility in uncertain construction projects. *Construction Management & Economics* 20(4): 343–352.

Ford, D. N., and Sobek, D. 2005. Modeling real options to switch among alternatives in product development. *IEEE Transactions on Engineering Management* 52(2): 1–11.

Garvin, M. J., and Cheah, C. Y. J. 2004. Valuation techniques for infrastructure investment decisions. *Construction Management & Economics* 22(5): 373–383.

Ho, S. P., and Liu, L. Y. 2003. How to evaluate and invest in emerging A/E/C technologies under uncertainty. *Journal of Construction Engineering and Management* 129(1): 16–24.

Howard, R. A., and Matheson, J. E. 1972. Risk-sensitive Markov design processes. *Management Science* 18(7): 356–370.

Johnson, S., Taylor, T., and Ford, D. 2006. Using system dynamics to extend real options use: Insights from the oil & gas industry. *2006 International System Dynamics Conference,* Nijmegen, the Netherlands, July 23–27.

Lander, D. M., and Pinches, G. E. 1998. Challenges to the practical implementation of modeling and valuing real options. *The Quarterly Review of Economics and Finance* 38:537–567.

Li, S. 2003. The role of expected value illustrated in decision-making under risk: Single-play versus multiple-play. *Journal of Risk Research* 6(2): 113–124.

Longstaff, F. A., and Schwartz, E. S. 2001. Valuing American options by simulation: A simple least-squares approach. *Review of Financial Studies* 14(1): 113–147.

McCullough, D. G. 1973. *The great bridge: The epic story of the building of the Brooklyn Bridge.* New York: Simon & Schuster.

Miller, R., and Lessard, D. 2000. *The strategic management of large engineering projects.* Cambridge, MA: MIT Press.

Moel, A., and Tufano, P. 1999. Bidding for the Antamina mine. In *Project flexibility, agency, and competition: New developments in the theory and application of real option analysis,* ed. M. J. Brennan and L. Trigeorgis, 128–150. New York: Oxford University Press.

Mun, J. 2005 *Real Options Analysis: Tools and Techniques for Valuing Strategic Investment and Decisions.* New York. John Wiley.

Ng, F., and Bjornsson, H. 2001. Evaluating a real option in material procurement. http://e-aec.stanford.edu/projects/docs/OptionPaper.pdf. Visited February 2, 2004.

Ng, F., Chiu, S., and Bjornsson, H. 2002. Quantifying price flexibility in material procurement as a real option. CIFE Technical Report no. 142. Palo Alto, CA: Stanford University.

Sabbagh, K. 1991. *Skyscraper: The making of a building.* New York: Penguin.

Schmidt, J. 2003. Real options and strategic decision-making summary. *Seminar in Business Strategy and International Business* TU-91.167.

Teach, E. 2003. Will real options take root? Why companies have been slow to adopt the valuation techniques. *CFO Magazine,* July.

Triantis, A. 2005. Realizing the potential of real options: Does theory meet practice? *Journal of Applied Corporate Finance* 17(2): 8–16.

Trigeorgis, L. 1993. Real options and interactions with financial flexibility. *Financial Management* 22(Autumn): 202–224.

Zhao, T., Sundararajan, S. K., and Tseng, C. L. 2004. Highway development decision-making under uncertainty: A real options approach. *Journal of Infrastructure Systems* 10(1): 23–32.

Zhao, T., and Tseng, C. 2003. Valuing flexibility in infrastructure expansion. *Journal of Infrastructure Systems* 9(3): 89–97.

8 Identifying Real Options to Improve the Design of Engineering Systems

Richard de Neufville
Massachusetts Institute of Technology

Olivier de Weck
Massachusetts Institute of Technology

Jijun Lin
Massachusetts Institute of Technology

Stefan Scholtes
University of Cambridge

CONTENTS

This chapter is part of the developing literature on the use of real options in the design and development of major projects. Chapter 9 by Kazakidis and Mayer and Chapter 10 by Kalligeros are part of this stream. The field is growing rapidly because engineering projects are expanding in scope and complexity and their environment is thus becoming increasingly uncertain. This chapter locates this work between standard engineering practice (which does not deal systematically with design flexibility) and financial real options analysis (which does not deal with system design).

The focus on flexibility in design thus involves deep reframing of the way we think about both design and the use of real options. It involves important changes in paradigm and practice.

This chapter deals with a core issue in the development of flexible designs: the problem of identifying the most desirable sources of flexibility. We present the use of screening models as an efficient procedure for solving this problem. We couple these models with simulations of the possible future outcomes driven by the range of values of factors that influence system performance. We illustrate the use of screening models using a case study of the development of a hypothetical deep-water oil field.

8.1 CONCEPT OF FLEXIBILITY IN DESIGN

The work on flexibility in design aims to improve the overall value of projects. The motivation for this effort lies in the fact that flexibility can provide truly significant increases in overall expected value, 30 percent or more in many instances, compared to standard designs that do not incorporate flexibility. By value we refer to output of the system, along whatever metric is relevant to its managers; this may be profit or net present value (NPV) for a commercial company, lives saved for a health care enterprise, or pollution reduction for an environmental agency. Conversely, we are not referring to the technical excellence of a system whose performance is considered a sine qua non.

Flexible designs can both provide substantial valuable insurance against downside risks, and enable system managers to take advantage of new opportunities. In general, flexibility in design shifts the distribution of outcomes from the less desirable to the more favorable. Both the case study presented here and numerous examples demonstrate these possibilities. See, for example, de Neufville et al. (2006), de Weck et al. (2004), and Hassan and de Neufville (2006).

Flexibility is most valuable when uncertainty is high, as is well-known from the financial literature on options. Flexibility in fact derives its value from uncertainty. If there were none, if we knew what would happen, there would be no point in having contingency plans or insurance. Conversely, the greater the risks, the more valuable insurance policies become. De Weck et al. (2007) indicate how this plays out in product design.

Flexibility is therefore most valuable for longer-term projects. This is for two reasons. First, the future is most uncertain for longer-term projects. As many studies consistently demonstrate, forecasts are "always wrong" in that the actual demands for services or prices of goods may easily be less than half or more than twice as much as anticipated (see, e.g., Flyvjberg et al. 2003, 2005). There is thus great virtue in waiting to find out what is necessary and to design accordingly.

Flexibility in design is further valuable because activities that last a long time give the system managers considerable scope to decide on the size and timing of investments, and thus to modify the cash flow to increase the NPV. The ability to expand a project easily, and as needed, makes it possible to defer investments and thus to discount their negative impact on present values, or to avoid investments altogether if future circumstances turn out to be unfavorable for expansion. If it is possible to expand both when and how necessary, then it is possible to avoid creating an initial project that embodies the capability to meet possible future "needs"—that may not ultimately be required.

Flexibility is thus valuable for all kinds of projects in which the designers can influence the size and timing of projects. It adds value to all kinds of large-scale infrastructure projects, such as highways (de Neufville et al. 2008a), railroads (Petkova 2007), airports (Chambers 2007), and communications networks (de Weck et al. 2004; Hassan and de Neufville 2006). Equally, it applies to projects in many different areas, such as the extractive industries (Babajide 2007), manufacturing and product platforms (Suh 2007), and hospital services (Lee 2007; de Neufville et al. 2008b). As shorthand for this range of interest, we usually refer to flexibility in system design.

Creating flexibility in design is equivalent to creating real options. For example, designing an aircraft so that the fuselage can easily be stretched to accommodate more passengers or cargo gives the manufacturer the "right, but not the obligation" to produce larger models. However, these options created in the design of the system differ from real options as commonly understood.

We refer to options that involve design features as "real options **in** systems" to distinguish them from real options that treat the technology as a "black box" and that can be thought of as "real options **on** systems" (Wang 2005). The option to acquire or open a mine, for example, does not involve any design issues. Most of the literature on real options refers to those "**on**" systems. The distinction between options "**on**" and "**in**" systems has meaningful conceptual and analytic consequences.

8.2   REAL OPTIONS "IN" SYSTEM DESIGN: A PARADIGM CHANGE

It is important to understand that the explicit consideration of flexibility in system design represents a considerable departure from current engineering practice. Although designers often promote the idea of flexibility, the fact is that the recognition of the need for and value of flexibility in design represents a paradigm change for many practitioners.

The focus on flexibility in system design represents a paradigm change because it presumes that the major requirements of the system are—at least partially—unknown and, indeed, unknowable in advance. The rationale for flexibility in design, for options in general, is precisely that the future is uncertain and that there is value in having the "right, but not the obligation" to react to future developments. The options to exit from a bad situation, or to take advantage of new opportunities, make sense only to the extent that one recognizes in advance that such situations may occur. However, the practice of engineering design routinely assumes that unknowable future uses of the system can be and are, in fact, known! To shift from the assumption that major design factors are known, to the reality that they are not, may entail a difficult conceptual and organizational break.

The empirical fact is that the design of important engineering systems routinely proceeds on the basis of fixed assumptions. For example:

- The design of the $5 billion Iridium system of communication satellites was based on the assumption made around 1990 that the system would have a subscriber base of 3 million customers once deployed in 1998. This was done despite the reality that the future demand for a new technology a decade hence is highly uncertain—and in this case the customer base at the time of bankruptcy of the system was only about 50,000 (de Weck et al. 2004).
- It is common practice in the extractive industries (petroleum and mining) to assume that the future price of the output product is fixed. Thus Codelco, the Chilean national copper company, assumed a long-term price for copper of about $1 a pound, when the actual price had fluctuated between $ 0.70 and $ 4.00 a pound within the previous five years (de Neufville 2006). Likewise, the major oil companies were in 2007 basing investment decisions on the notion that future prices were fixed at about $35 a barrel (the exact assumed amount varies among companies and is highly confidential) in the context of actual prices varying by over a factor of 4.
- The design of U.S. military systems is typically based on "requirements," despite the continuing experience that the actual role of military systems once deployed often differs greatly from that originally anticipated (Bartolomei 2007).

The engineering tradition of assuming that major design requirements can be known appears to have evolved from two complementary forces:

1. The notion that many important drivers of value—such as the demand for satellite communications services or the future price of oil or copper—are social factors outside of the purview of engineering
2. The very complexity of design, which impels designers to make simplifying assumptions, to be able to execute the kind of detailed design that they ultimately have to make in order to ensure technical feasibility

It certainly is true that in the extractive industries, the financial sides of companies insist on setting the price of the product (oil, copper, etc.) that designers will use when they evaluate alternative designs. An important rationale for this approach is to establish a level playing field for the promoters of projects throughout the company. The downside of this approach is of course that if you assume that the price of oil will be $35 a barrel for the life of the project, you will not make any provisions for extracting oil from fields that make sense only when the price is higher than $35 a barrel. By making such deterministic assumptions, one may thus leave considerable value unexploited.

In any case, the practice of disregarding variability in future prices of products is firmly embedded in many corporate cultures. Many engineers feel that this situation is in the natural order of things, as they should stick to technology and leave economics to others. Many senior engineers in these companies also consider that it is taboo to consider price variability in design analyses. Even when designers accept the desirability of recognizing the variability of price and exploring how this can lead to improved designs, they face tough organizational battles with the financial side of the company, which naturally does not want to give up the power it has been exercising over the designers.

Purely within the context of engineering practice, moreover, it has traditionally been felt necessary to simplify the design problem by assuming that important parameters are fixed. To appreciate this, consider Figure 8.1. It first lays out the full scope of what it would take to analyze the entire planning, design, and operational management of a system. As indicated by Cardin et al. (2008), complete examination of all the possibilities would consider:

- All the combinations of the design
- Crossed with all the ways that outside factors (such as price and level of demand) would influence its performance
- Further crossed with all the possible ways that managers might react to situational variations
- Considering a wide range of metrics of interest to the several stakeholders for the system

This problem is computationally intractable now, as it will be for a long time. For example, if we conservatively assume that the design space for a major system allows for 1,000,000 different combinations of the settings of the design variables, that there are 10,000 different demand scenarios, 100 different decision rules, and 10 metrics of interest, then this leads to a combinatorial space of order $O(10^{13})$. To make progress, the design problem must be simplified.

Current practice examines only a portion of the total design problem. As Figure 8.1 indicates, the analysis process typically assumes fixed values for important factors.

Situation	Initial Design	Outside Factors	Managers Adjust	Metrics Used
Complete process for development of a complex system	Physical infrastructure	Price, demand for services	Best use of existing facilities; development of additional facilities	Net present value, rate of return, initial capital expenditure, etc.
	(many possibilities)	(many possibilities)	(many possibilities)	(many possibilities)
Current practice	Physical infrastructure	Price, demand for services	Best use of existing facilities; development of additional facilities	Net present value, rate of return, etc.
	(A few possibilities)	(1 value for each)	(1 plan)	(1 cash flow)

FIGURE 8.1 Detailed contrast between real situation and current design practice.

	Initial Design	Outside Factors	Managers Adjust	Performance
Actual situation	Many possibilities	Many possibilities	Many possibilities	Many possibilities
Current practice	A few possibilities	(1 value for each)	(1 plan)	(1 cash flow)

FIGURE 8.2 Summary contrast between real situation and current design practice.

The logic of the situation is then that it considers one implementation plan, and makes no allowance for the reality that the system owners will manage the asset intelligently and react to circumstances as they evolve—an assumption that of course is contrary to what actually happens. Finally, this analysis leads to a single cash flow associated with any design and a single set of equivalent measures of the discounted value of that unique cash flow (such as NPV, internal rate of return, or return on investment).

Carrying out this analysis properly in detail for any major system (such as for a fleet of satellites, a hospital, or an oil platform) is a major project. Engineers who have spent years on this kind of evaluation are naturally reluctant to consider expanding their large task. Yet there is a great gap between current practice and the reality of what needs to be done for complete system planning, design, and management, as Figure 8.2 indicates. Bridging this gap will indeed be a paradigm change in engineering practice.

8.3   FOCUS ON IMPROVED DESIGN, NOT ON PRICE OF OPTION

The root issue in creating a flexible design is to determine which part of the system should be flexible. Which parts should be configured to provide the real options that will give the system managers the "right, but not the obligation" to change the size or function of a system?

Determining where flexibility should be embedded in a design is an immense issue. Even fairly simple systems involve many different elements that could be designed to be adjustable. Consider the design of unmanned aerial vehicles (UAVs). These are bigger, more sophisticated versions of radio-controlled model aircraft. They have six major parts: the wings, the fuselage, the tail, the motor, the control element, and a payload. Flexibility could be designed into each of these elements, and could be useful in different circumstances, depending on the future possible use of the UAV. Identifying where to embed flexibility in the UAV is not obvious, and requires careful analysis (Bartolomei 2007). For larger systems, such as a communications network, an automobile production process, or a refinery, the number of combinations of design parameters can be astronomical. Further complicating the problem, the determination of what real options should be developed requires that we consider how the many possible designs perform under a range of possible scenarios. The total number of possibilities can be staggering. For example, Hassan and De Neufville (2006) estimated that the design space for her very simplified model of the exploitation of an oilfield involved 10^{60} possibilities.

The size of the problem involved in determining which real options should be built in has important consequences for the valuation of the options. It means that the valuation process used in this analysis cannot be fully exhaustive and accurate. Refined, final valuations have to be done in a different process.

The logic behind this conclusion is straightforward:

1. The size of the problem involved in determining what flexibility should be involved in a design is so large that
   (a) Simplified methods need to be used in order to explore the possibilities.
   (b) The associated estimates of the value of the possible design cannot be precise—they are indicative, they represent a sample taken from a larger set, and they are not final.

It is to be stressed that the analysis to determine design flexibility is aimed at only the up-front stage of the design process. Indeed, the process of identifying the design elements whose flexibility most strongly increases expected value is not the same as the process of designing the system in detail. On the contrary, the overall design process inherently involves two stages:

1. Some process for defining candidate overall configurations of the system and the extent and nature of their design flexibility. This stage is typically referred to as "conceptual design" or "systems architecting".
2. Detailed analyses of these architectures to determine their exact designs and to evaluate them in detail to select one for implementation. This is the "detailed design" or "define" stage.

An example illustrates the point. In designing parking garages for a growing area, a critical question is: How many levels should it have? An analysis of design flexibility for a particular location—only possible using an approximate model of the construction costs and prospective revenues—convincingly indicated the desirability to have the flexibility to increase the number of floors, in case demand grew to justify them. The flexible design in this case increased expected value by about 25 percent, reduced initial capital expenditures by about 20 percent, and both reduced the value-at-risk (VAR) and increased the maximum possible benefits (de Neufville et al. 2006). The valuation process used was crude, but sufficient to produce the result. The numbers generated to achieve this useful result are not, however, sufficient to provide the kind of refined, final valuation that will ultimately be needed to justify the financing of the project.

In short, the roles and tasks of system designers are different from those of financial and real options analysts. Systems designers identify which architectures offer the greatest possibilities and can use simple, approximate valuations to achieve good results. This phase defines a few possible designs with varying degrees of flexibility (real options content). Economic analysts justifying budget allocations to a final design need the much more detailed designs developed from the optimal overall architectures.

Overall, as the preceding sections indicate, the work on design flexibility lies between

- Financial real options analysis—which does not deal with system design
- Standard engineering practice—which does not deal systematically with design flexibility

In short, dealing effectively with design flexibility involves a much larger view of the design process, but a simplified, more approximate take on the valuation of the options.

8.4 IDENTIFYING BEST OPPORTUNITIES FOR FLEXIBILITY

The identification of the best opportunities for flexibility requires us to apply new measures of value. Just as the work is situated between standard engineering practice and financial real options analysis, so must our metrics be. Thus,

- From the perspective of standard engineering, the metrics need to reflect the reality that future performance is a distribution of possibilities, and cannot be fairly encapsulated by any single deterministic measure such as NPV.
- From the perspective of financial real options analysis, it is unrealistic to assume that the measures of value that can be generated from analysis of a design can in any way reflect a proper risk-balanced market price, enforced by the possibility of arbitrage pricing.

In sum, our measures of value need to be more complete than those commonly used in engineering, and more practical and less theoretical than those appropriate to sophisticated options analysis rooted in the financial derivative markets.

Once we recognize that there are uncertainty and a distribution of possibilities, it is no longer meaningful to think of a deterministic measure of value of a system, such as a discounted cash flow metric like NPV. At the very least, we need some comprehensive view of the possible values associated with a design. The expected net present value (ENPV) is the more obvious of the possible summary measures, and probably the easiest to understand. As the name implies, the ENPV is simply the average discounted value of a design.

A single measure is not enough, however. It is easy to imagine two sets of distributions that share one measure but differ markedly from each other. For example, two distributions might have the same ENPV, but one could have a much narrower standard deviation. A person or organization might then prefer the distribution with the narrower distribution—and less risk. Thus, a single measure cannot fully inform the decision-making process.

Nor can we reasonably argue that one measure is necessarily better than others. Although it might be tempting to think of ENPV as the primary measure of value—by analogy with NPV—it could easily be misleading. Decision makers might prefer a design with a smaller standard deviation (taking it to be less risky) even though it has a lower ENPV. In short, we need to think of ways to consider the whole distribution of possible outcomes.

The value-at-risk-and-gain (VARG) diagram is a convenient way to display the distribution of possible results. It graphs the cumulative value associated with any possible design. It builds upon the VAR concept used by bankers to identify the risk of the losses they might incur. As might be expected, lenders focus on the likelihood that they might not recover their loans. Investors, however, are concerned not only with possible losses but also and most importantly with the amounts they might gain. Thus, the VARG curve displays the entire range.

We arrive at the VARG in two steps:

1. For each potential design (with or without embedded flexibility), uncertain future scenario, and managerial decision making, we evaluate the outcomes over time. If our metric is profits, we consider possible future cash flow and develop an NPV.
2. We then consider the distribution of possible scenarios, each representing different demand scenarios and managerial responses, to obtain a distribution of NPVs. This is conventionally shown in the form of a cumulative distribution function, as in Figure 8.3.

Many metrics can be read from a VARG curve. The ones that seem most useful to decision makers are the EPNV, and the maximum and minimum results. It is also possible to calculate the standard deviation, which measures the spread of the distribution. Many designers use this measure to calculate the "robustness" of a system, that is, the degree to which its performance is affected by uncertainties. In some cases (such as the tuning of a radio), this is a good feature. In general, however, managers prefer designs skewed toward upside gains, thus with high upside spreads.

Initial capital expenditure (CAPEX) is another useful measure of performance, in addition to the ENPV and the VARG. Initial CAPEX is a particularly interesting metric because it is often a main element distinguishing a conventional inflexible design and one with options. This is because designs with options are generally staged (representing an initial implementation followed by later stages in which options are possibly implemented). Staged developments correspondingly are cheaper initially than conventional designs, because they defer some build-outs until future events

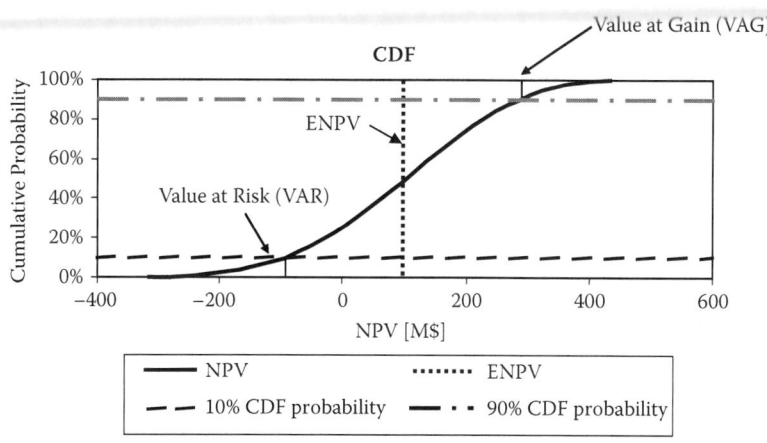

FIGURE 8.3 Value-at-risk-and-gain (VARG) curves for a hypothetical project.

demonstrate the need for expansion. As can easily be appreciated, a design with a lower level of initial CAPEX leads to higher returns on investment for the same amount of EPNV as a design with a higher initial CAPEX. Thus, the level of initial CAPEX is often a primary distinction between designs with various levels or forms of flexibility.

Decision makers may be interested in the probability of levels of minimum and maximum returns from a project. For example, one might want to know the possible downside with a 10 percent chance of occurrence, which is known as "10 percent value at risk (or 10 percent VAR)." In Figure 8.3, this is $90 million. Conversely, one might focus on the upside potential or "value at gain" (VAG). For example, the 10 percent VAG (corresponding to the complementary 90 percent VAR) is read at the 90 percent level of the vertical axis. Figure 8.3 indicates that there is a 10 percent chance that the gains from the system will be greater than $290 million.

Other measures are possible. The important point is that once we deal with distributions of outcomes, we need a range of measures of the value a system may deliver. As indicated in the case study presented later on, a combination of graphs and tables provides a convenient way to represent the measures of value for a system with various degrees of flexibility and to identify those that should be examined in detail.

Overall, we use the VARG graphs, complemented by a table summarizing the several metrics of value, to help identify which design architectures are most attractive for detailed design. When higher values are toward the right, as in Figure 8.3, then configurations with curves and ENPV toward the right are better. The better alternative is obvious if its curve lies entirely to the right of another. In general, however, trade-offs have to be made since it is rare that one design absolutely dominates all others.

A key point is that as system designers, we have the ability to improve designs that have uncertain outcomes. We do this by shifting the VARG curves to the right as much as possible. By intelligently screening for, designing in, and exercising flexibility in the form of real options throughout a system's life cycle, we may in fact "design" or at least influence the shape of the NPV distributions given a set of exogenous uncertainties. We do this typically by two complementary types of actions: one set reduces the downside tail (acting as "puts" for the system), and the other set extends the upside tail (acting as "calls"). Overall, we want to make the ENPV higher.

8.5   USING SCREENING MODELS TO IDENTIFY VALUABLE FLEXIBILITY

8.5.1   ANALYTIC APPROACH TO SCREENING MODELS

A proper search for the best configuration of a system cannot be done in great detail. Explicitly or implicitly, the investigation needs to consider many thousands of possibilities, as Figure 8.1 suggests. Whether the focus is on a network of communications satellites, a deep-water platform for exploiting petroleum reservoirs, or a hospital, the range of possible design elements and the way they can be combined is very large.

The standard engineering approach to design focuses on the detailed definition of the possibilities. This is natural, since the engineering team ultimately has to specify the design in great detail. The design process thus revolves around one or more very detailed—"high-fidelity"—models of the technical aspects of the systems. Using these high-fidelity models to create a single design can easily take months—a single run of a detailed model of a system may take a day or more.

It is simply not practical to use fully detailed, high-fidelity models to determine the best configuration of a system in the context of major uncertainties. Any proper analysis of possible scenarios necessitates the examination of a large number of possible designs: 500 is a standard number at present. Exploratory analyses of alternative configurations for a system cannot afford to carry out 500 individual trials that each takes a day or a week.

To conduct a proper search for the best configuration of a system, it is necessary to use models that can be run very much faster than the detailed, high-fidelity models. To achieve this speed, it is necessary to sacrifice some of the detail. As Figure 8.4 suggests, the analysis needs to trade details for speed. As indicated in the case study presented in Section 8.5.3, it is frequently possible to define reasonable mid-fidelity models of a system that run about 1,000 times faster than the completely detailed high-fidelity models. Correspondingly, this permits the consideration of about 1,000 cases instead of only one within the same time.

Mid-fidelity models used for the rapid exploration of a complex system are known as "screening models." The name indicates their use. Screening models examine many different designs to identify—to screen out—the smaller number that deserve detailed consideration. Screening models have been used ever since the development of optimization techniques. Jacoby and Loucks (1972) provide a classic example of the use of screening models in the design of infrastructure systems, and de Neufville and Marks (1974) compile others. Wang (2005) and de Neufville, (2006) extended the use of screening models to the analysis of flexibility in systems design.

Screening models identify system configurations that are most worthy of detailed analysis and evaluation. They are properly used in early parts of the analysis, as

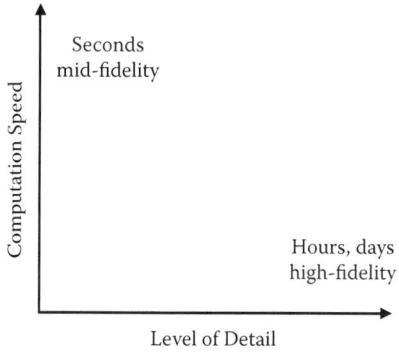

FIGURE 8.4 Trade-off between high-fidelity models taking days or weeks to run, and mid-fidelity models that run in seconds or a few minutes.

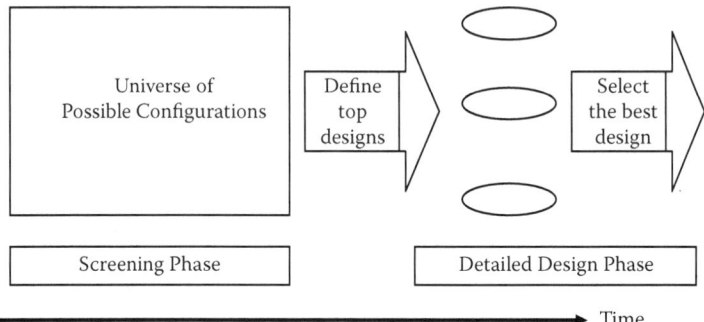

FIGURE 8.5 Screening models help define top designs for detailed analysis and design.

Figure 8.5 suggests. Screening models enable the designers to sort through the universe of possible system architectures rapidly, and to define top designs.

Screening models provide an analytic process for determining which system configurations are potentially most valuable. They complement the often intuitive ways that engineers now use to fix on the architectures they will examine in detail. Indeed, it is often the case that the design team fixes on possible system architectures based on the intuition or experience of the lead engineers, on what some competitors have recently done, or on what the team perceives to be good practice. Screening models provide a much more rigorous, more analytical—and more of an engineering—approach to the definition of which designs the team should analyze in detail. As Figure 8.5 suggests, they provide a transition between the large universe of possibilities and the narrow focus on a few possible designs.

Note that the use of screening models does not have to be fully automated. In some cases, the range of designs can be fully described analytically and explored through an automated process. Alternatively, as in our case study, the process is hand-tailored by the analyst. In this case, the systems designer—guided by experience, intuition, or the direction of the client—picks a few trial configurations to be explored and tries them out. The initial results guide the subsequent choice of trial configurations.

Screening models work in combination with detailed design models. Each type fulfills its distinct role. The screening models operate at the first stage of the design process; they identify overall design configurations for detailed design. Complementarily, the detailed design models refine the configurations suggested by the screening models. As such, each type needs to be judged on its own merits for fulfilling its role. The screening models should not be judged on the basis of their deficiencies compared to the detailed models, and vice versa.

Screening models need to be judged on the basis of their effectiveness in correctly identifying interesting configurations for detailed consideration. Not just any simplified model constitutes an effective screening model. A priori, it is never obvious which simplifications of the detailed models are appropriate and useful. Proposed screening models need to be validated. Generally speaking, screening models should be accurate enough to reflect the *relative* NPV rankings of different designs under uncertainty. Detailed models, on the other hand, focus on producing answers that are accurate on an *absolute* scale.

8.5.2   SCREENING FOR REAL OPTIONS

Screening models enable the transition from current practice to a more comprehensive analysis of projects. Because these mid-fidelity models are faster than the high-fidelity models used for design, they reduce the computation effort required for the consideration of the initial design, and allow for a larger analysis of the effect of outside factors, such as variations in the price of or demand for the product. Figure 8.6 illustrates the possibility.

Of course, we want to make sure that the use of the mid-fidelity models is effective, that is, that they guide us correctly to interesting designs, with a minimum number of false positives. Thus we need to vet the mid-fidelity models. We need to verify that they correctly describe the physical realities, as regards not only the detailed elements (the "trees") but also the overall features, such as the existence and degree of economies of scale (the "forest"). Functionally, we need to see to what extent they correctly rank alternative designs—even when the absolute values they assign to alternatives are approximate.

Using screening models to identify real options is novel. Known previous applications of screening models have assumed that the major specifications for the system were fixed, and the screening models were directed toward breaking down the analysis of the design into computationally manageable bites. The work of Wang and de Neufville (2006) possibly represents the first application of screening models to the identification and definition of opportunities for flexibility and real options.

The use of screening models for identifying real options requires a major intellectual jump for many engineers. Because options are valuable only when there is uncertainty, the screening models used to identify real options must explore the uncertainties that surround a system, such as the future price or quality of a product, the demand for it compared to new technologies, and the regulatory or political context. The analysis has to consider the distribution of these uncertainties and outcomes. This fact naturally complicates the analysis compared to the conventional deterministic analysis: each run of the model must consider and compare many possible outcomes instead of only one. This complicates the evaluation and comparison of the choices: we require a new set of criteria and procedures for choosing the best designs.

	Initial Design	Outside Factors	Managers Adjust	Performance
Actual situation	Many possibilities	Many possibilities	Many possibilities	Many possibilities
Using screening models	Many possibilities	Some important possibilities	Some possibilities	Many cash flows
Current practice	A few possibilities	1 value for each	1 plan	1 cash flow

FIGURE 8.6 Contrast between possibilities using screening models and current design practice.

From a purely analytic perspective, however, using screening models to identify real options is straightforward. It consists of applying the range of possible outside factors to the screening model, the simpler, midfidelity representation of an initial design. This process can be automated by simulating the assumed distributions of the outside factors. Such a simulation is easily implemented as an add-in to a spreadsheet, or can be executed using any of a number of commercial packages, such as Crystal Ball®.

8.5.3 EXAMPLE PROBLEM

The case study presented in this analysis was inspired by the issues concerning the design of deep-sea platforms for the exploitation of oil fields. These are massive, highly expensive long-term projects. The platforms easily cost from $300 million to $1 billion each, even when in relatively easy conditions (such as off-shore of Angola) and easily much more when they have to deal with especially challenging conditions (such as in the Arctic sea ice).

These projects exist in highly uncertain environments. Great uncertainty inevitably surrounds the

- Quantity of oil that may be extracted: this depends on the detailed geological characteristics of the substructure, which will not be fully known until long after the project starts (see Babajide 2007).
- Quality of the oil, that is, the degree of oil and gas, the hydrocarbon composition including sulfur and water contents.
- Value of the oil to the investor: determined globally by the highly volatile world market, and further by the revenue-sharing agreements with the owners of the oil fields. As a matter of record, the price of oil ranged from $15 to almost $150 per barrel between 1990 and 2008.

Given the great uncertainty inherent in these projects, they are prime candidates for the use of flexibility in design, whose value is greatest when the uncertainties are highest.

8.6 DEVELOPING SCREENING MODEL

The designers of oil platforms use highly sophisticated models for each of the major portions of the platform system. Generally, as Figure 8.7 indicates, systems consist of models of the

- Inputs, or the nature of the oil field
- Production, or the possible physical design of the platform, its wells, and the connections between them
- Output, immediately in terms of the oil and gas produced, but most significantly in terms of economic value, which places a premium on immediate gains and discounts future revenues.

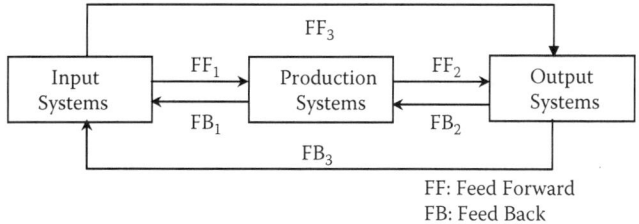

FF: Feed Forward
FB: Feed Back

FIGURE 8.7 A generic integrated systems model of uncertainty in petroleum projects.

Complex feed-forward and feed-back mechanisms connect these major elements. They relate to the

- Physical flows of materials (oil, gas, water, etc.) going through the system
- Logical flows that capture the control logic that regulates the physical flows, such as the rates, timing, and capacity constraints on physical flows
- Financial flows, generally driven by the physical flows in terms of revenues, and by the investments and operations in terms of costs

For example, the design of the production system affects the output, but the output over time defines what kinds of inputs—such as water or gas injections to sustain pressure—will be needed.

The high-fidelity versions of these models are generally distinct and separate. This is because they are usually developed by very different groups of professionals, ranging from the petroleum geologists who characterize the oil field, through the engineers who design the platform and wells, to the economists who model the financial performance. In the context of high-fidelity design, these models are run separately, and do not exchange data or "talk" to each other each time they are run. This arrangement is satisfactory in the context of detailed design.

A screening model must be integrated, however. It needs to bring together the distinct phases of the detailed design. This is because we will use it to analyze thousands of situations, defined by the crossing of various trial designs with a wide range of possible scenarios. Developing an integrated, mid-fidelity model from the high-fidelity versions thus requires considerable effort—easily on the order of six person-months.

Screening models also need to be perfected and validated for use without overburdening them with excessive detail. What does and does not represent necessary detail in a screening model is not an easy question to answer. In general, the screening model should include specific factors that may significantly affect the value metrics (such as NPV and VARG elements) and the rank order of the alternative system architectures. But since screening models are needed to produce VARG estimates for designs, there is a chicken-and-egg problem with respect to their fidelity. The issue can be resolved by taking an iterative approach. A screening model will have sufficient fidelity when it meets acceptable criteria, for example when the iterations rank designs (with varying degrees of flexibility) consistently and the VARG estimates stabilize to within 5–10 percent compared to earlier estimations. The specific details to be included in a screening model will depend on the application.

Offshore oil and gas projects face both large business risks and significant opportunities. A major source of uncertainty concerns the volumes and composition of the hydrocarbon reservoirs in a field. Another relates to the market environment, particularly the oil and gas prices. Both these internal and external uncertainties have significant impact on the success of projects and need to be taken into account when defining the strategies for developing a project, most specifically by building in the flexibility to deal with the alternative scenarios they may present.

Estimates of the size of an oil field may vary greatly. The oil industry usually defines this in terms of the stock tank original oil in place (STOOIP). From one perspective, this is a fixed number determined by long-term geologic conditions. However, as a practical matter it is an estimate likely to have a wide distribution. The STOOIP for a field normally changes considerably over the life of a project, as further exploration and then exploitation of the field proceed, and as more information about its details emerges. Moreover, the estimated STOOIP may not converge to a single number. New discoveries of reservoir conditions, such as its degree of fragmentation, may disruptively change the estimate. As the STOOIP is a main driver of many design decisions, it is very important for decision makers to recognize reservoir uncertainty and take it into account in their design of field development strategies.

As the historical record clearly demonstrates, the demand and prices for crude oil and gas are highly volatile. These variations strongly affect the success of field development and operations. For example, when oil prices and demand are high, early oil to market in the field will take advantage of the favorable market conditions; and thus a flexible design that can rapidly expand the capacity of a platform would be desirable. Decision makers need to consider the market uncertainties in defining their strategies for developing oil fields.

8.6.1 FLEXIBILITY IN PETROLEUM PROJECTS

Real options can be created at three levels of offshore petroleum projects. The flexibility can be

- Interfacility: considering how the major elements or facilities in a field might be linked. This is the highest level of flexibility, which considers that platforms and major subsurface elements can be added, moved, or retired from the field, or new reservoirs are connected to existing facilities over time. This kind of flexibility defines the topology choices between reservoirs and platforms. Typical examples are flexible staged development for a single large oil field or tie-back of a new reservoir to an existing platform.
- Intrafacility: applies within one field or one facility. It defines the design options of individual platforms, wells, and so on. Examples of such flexibility are adding extra space in the production, drilling, or cellar deck allowing the later addition of modules, such as the addition of more water injection pumps

or compression trains, or the decommissioning of a flare tower once gas export via pipeline to an onshore liquefied natural gas (LNG) facility begins.

• Operational: does not require changes to the physical system. For example, to achieve higher oil recovery rates from a reservoir, field operators can actively manage production by increasing water and gas injection rates, changing the mix of incoming fluids from different reservoirs, switching between production and injection wells, or temporarily shutting down wells.

8.6.2   STEPWISE DEVELOPMENT OF ANALYSIS

The analysis for the example oil field was carried out in steps, as Figure 8.8 indicates. This was done to help explain the use and significant value of screening models to identify opportunities to improve the design of the oil platforms. Once the use of screening models is widely understood, it will not be necessary to break the analysis out into these steps.

Step (1) represents traditional practice. It applies the screening model to a conventional design that optimizes project value based on a deterministic "best-guess" estimation of reservoir volumes and fixed oil and gas prices. It estimates a single NPV.

Step (2) evaluates the conventional design recognizing both reservoir and market uncertainty. It simulates the joint distribution of these factors and calculates the NPV associated with each sample and then the average or expected NPV. Note carefully that the ENPV in general differs from the deterministic NPV. This is due to Jensen's inequality (Wikipedia 2008), the rule stating that the true value given by the expectation over the possible states is generally not the value calculated from an expected value of a parameter unless all functions are linear.

$$ENPV = EV\,[f(x)] \neq f[\,EV(x)]$$

Step (2) presents results both in a table and using the VARG curve, which gives a more comprehensive evaluation of the project under uncertain factors. This step makes two points:

1. Ignoring the distribution of uncertain variables leads to an incorrect assessment of the NPV.
2. A project leads to a distribution of possible outcomes.

Steps (3) and (4) represent the screening model in action. They consider different kinds of flexibility in field development. Each possibility leads to a set of metrics that can be used to rank the possible developments, most importantly in comparison with the conventional design evaluated in Step (2). The result of this process is a short list of the best candidates for detailed design. The process thus identifies the kinds of flexibility—the real options—that seem to improve the overall performance of the design most significantly. This is the objective of the screening process.

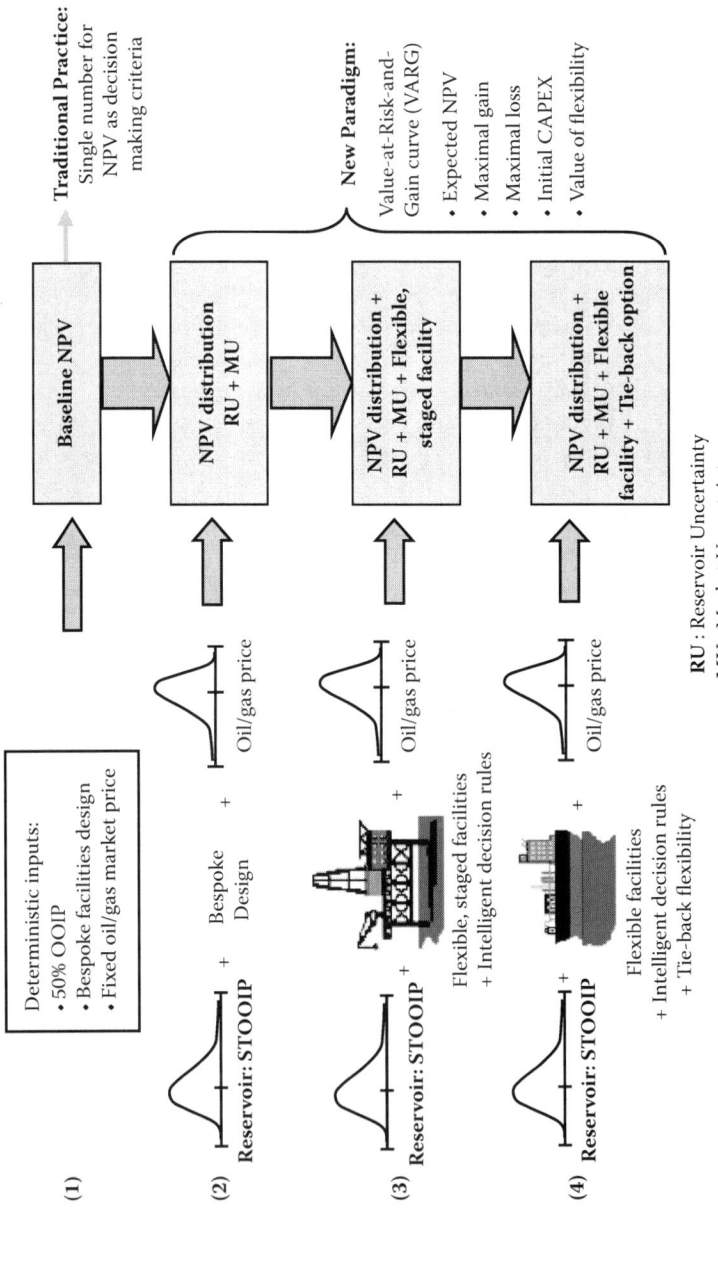

FIGURE 8.8 Applications of the integrated screening model to oil and gas field development.

8.6.3 ALTERNATIVES CONSIDERED

The analysis focused on the concept of modular design, the idea of building a complex system as a series of equal modules. This approach to system design is now being widely developed in the oil and gas industry, where it goes by the name of "design one, build many" (see, e.g., ExxonMobil 2008; The Peninsula 2007; Shell Technology and Innovation 2008).

The three alternative designs considered were

1. *A big monolithic facility:* This represents conventional design—a facility optimized to deal with the best estimate of the size of the field and pre-defined oil prices. Due to economies of scale, this design may benefit from big orders of equipment from suppliers. However, it may lead to oversized capacity if the reservoir and market conditions are not favorable.
2. *Predetermined staged development:* This strategy develops the facility in predetermined stages over time. One advantage of this strategy is that following stages can benefit from learning. Another is that it phases major capital investments over time.
3. *Flexible staged development:* This strategy allows the number and timing of stages to be flexible. The number of stages depends on the time-varying estimate of STOOIP. To analyze this case, it is necessary to posit a decision rule, that is, criteria that determine the timing and size of additional stages.

8.6.4 EXAMPLE CALCULATIONS

The analysis evaluated the three field development strategies under reservoir uncertainty. It used hypothetical but generally representative numbers. The results are purely illustrative. A Monte Carlo simulation of 600 samples represented the distribution of the STOOIP. Using a mid-fidelity integrated screening model, each run took only about 1.2 seconds. We could thus simulate hundreds of runs quickly. In contrast, the applicable high-fidelity reservoir models can take a couple hours for a single run.

Figure 8.9 and Table 8.1 display the results. They illustrate several important conclusions:

- The conventional design has considerable risks. It is attractive overall (ENPV = $6.15 billion). However, it has considerable downside risk if the quantity of oil and gas is low (minimum NPV ~ $1.3 billion loss). Moreover, it is limited on the upside in that it cannot expand capacity should this be desirable. Thus, the conventional design is not protected on the downside and cannot seize upside opportunities.
- The inflexible staged development has no overall benefit in this example. It reduces the present value of costs by deferring some investments and incorporating learning into subsequent stages. However, it increases costs by forgoing the economies of scale associated with a single large facility. Moreover, by postponing investments it also delays revenues and thus their

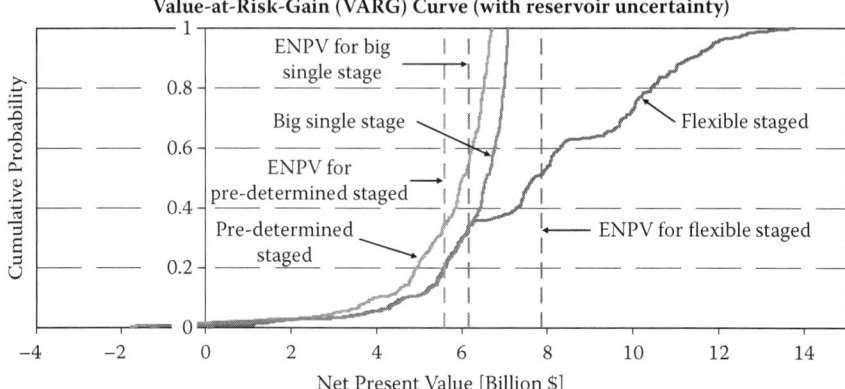

FIGURE 8.9 Value-at-risk-and-gain (VARG) curves for three field development strategies.

present value. The overall effect in this hypothetical case is to reduce the ENPV. Such an alternative offering no additional value should be "screened out" of further consideration.

- The flexible staged strategy offers significant improvements in this case. Because it adds stages only when the STOOIP is proven, it avoids potentially big losses. Thus its minimum NPV is much greater than that of a single monolithic facility. As it can be expanded beyond the capability of the single large facility, it can accelerate the recovery of revenues and achieve much higher NPV if the amount of oil is particularly high. These effects combine to create an ENPV of $7.86 billion or over 25 percent better than the conventional design. This strategy should thus be retained for detailed analysis.

Overall, this analysis supports the "design one, build many" strategy promoted by some companies. The display of the several metrics (Table 8.1) and the VARG curves (Figure 8.9) provide interesting insights into the benefits of suitable flexibility:

- They display the significant gains to be obtained by being ready to take advantage of upside opportunities. The flexible staged approach makes it possible

TABLE 8.1

Measures of Value for Development Strategies ($, Billions)

Development Type	NPV			CAPEX	
	Expected	Minimum	Maximum	Initial	Possible Maximum
Flexible staged	7.86	0.05	13.87	1.38	3.08
Predetermined stage	5.60	−1.77	6.72	1.40	1.40
Big single stage	6.15	−1.29	7.08	3.12	3.12

to achieve great returns in this particular example—more than double the average value of the conventional design—under favorable circumstances.

- Conversely, as is especially obvious in Table 8.1 but masked in the VARG graph, the flexible staged design avoids downside losses. Full implementation of this project can be stopped under unfavorable circumstances, and the potential $1.77 billion loss of the conventional design is essentially avoided.

- The flexible staged design almost doubles the rate of return. It requires only an initial investment of $1.38 billion to yield an ENPV of $7.86 billion, whereas the conventional design requires a more than double up-front commitment of $3.12 billion to obtain a lower overall ENPV of only $6.15 billion.

In short, the example analysis shows how using a screening model can identify design flexibility—real options embedded in the design—that can significantly improve the expected value of the project. The example shows the principle—it does not pretend to have considered anywhere near all the possibilities for flexible design, and thus does not pretend to have identified the absolute best.

Further, the example shows that even a relatively simple first-pass analysis that recognizes uncertainty can lead to flexible designs that offer impressive—in this case, more than 25 percent—improvements in value over conventional approaches.

8.7   SUMMARY AND FUTURE WORK

Research in real options is leading to significant practical results. Many applications demonstrate that flexible designs can greatly increase expected overall performance compared to traditional rigid designs optimized for specific conditions. Flexible designs can deliver these benefits because their underlying architecture enables managers to adapt projects to circumstances that develop. Owners can thus cut losses by avoiding undesirable outcomes, and increase gains by taking advantage of new opportunities.

This chapter focuses on the development of valuable flexibility in designs. It is part of a growing field of practice that promises—and so far demonstrates—great potential for significantly improving design practice. To give it a name, it is about "flexibility or real options **in** systems."

Conceptually and professionally, this work lies midway between standard engineering (which does not consider design flexibility in any detail) and financial real options analysis (which does not look at design). This position requires a considerable professional jump for engineering practitioners on the one hand (who need to engage explicitly with uncertainties in the design requirements), and for economic analysts (who in this context need to consider broader valuation metrics and to work more closely with their engineering counterparts) on the other.

The chapter specifically presents the "screening model" approach to the core problem of identifying the system elements that should be flexible in order to increase value. Screening models are mid-fidelity models that run much faster than standard detailed design models. They can examine the performance of many designs across great ranges of scenarios, and can thus identify system architectures that are most attractive as prospects for detailed design.

The case study illustrating the use of screening models to identify desirable flexible designs also suggests how these can be enormously valuable. In this case, the possible increase in ENPV is about 25 percent! This benefit is furthermore accompanied by a decrease in the risk of losses and increase in the possible gains!! It furthermore reduces the initial CAPEX investment, and thus substantially boosts the return on this investment!!!

The case study also suggests two areas of future work:

1. It is important to develop criteria for the scope and level of detail in the screening model required to produce reliable results. What level of detail should be included in the different input, output, and production system components of the screening model? How should it account for the various economic feed-forward and feedback mechanisms? These can be significant, as experience between 2006 and 2008 indicated. As the oil price jumped (from $30 to nearly $150 per barrel), so did the costs of raw materials, equipment, and contractor services. The cost of drilling wells more than doubled. Feedback between oil price and the cost of contractor services and equipment eventually impacts the capital and operating expenditures. When such phenomena affect the rank order of alternative designs, they should be included in the screening model. The issues concerning the fidelity of screening models need to be formalized in the future.

2. We also need to deepen our understanding of how and why changes to existing engineering systems happen. How are changes initiated, and how do they propagate through the system and cause other "knock-on" changes (Giffin et al. 2007)? The case study assumed that managers would be entirely "locked in" to the single facility even if the amount of oil was much larger than anticipated. In reality, this is not so. At some point, managers would expand the monolithic platform, albeit in a reactive mode—slower and more expensive than if the possibility of expansion had been designed into the system. A more refined screening model may therefore include the possibility and cost of making such reactive changes and might consist of an ensemble of discrete change cost reducers (Silver and de Weck 2007). Since designers and engineers are intimately familiar with the need and difficulty of making changes to their configurations, this interpretation of real options may help to further anchor the concept of flexibility in system design in modern engineering practice.

ACKNOWLEDGMENTS

The ideas presented here build upon a generation of work in the development of procedures for the effective planning, design, and implementation of engineering systems. Most recently, it has benefited from the support of BP, the Cambridge-MIT Institute, General Motors, Laing O'Rourke plc, the MIT-Portugal Program, the MITRE Corporation, and the U.S. Air Force.

REFERENCES

Babajide, A. 2007. Real options analysis as a decision tool in oil field developments. SM thesis, MIT, Engineering Systems Division. http://ardent.mit.edu/real_options/Real_opts_papers/Babajide_Thesis_FINAL.pdf.

Bartolomei, J. 2007. Qualitative knowledge construction for engineering systems: Extending the design structure matrix methods in scope and perspective. PhD dissertation, MIT, Engineering Systems Division. http://ardent.mit.edu/real_options/Real_options_papers/Bartolomei%20Thesis.pdf.

Cardin, M.-A., de Neufville, R., and Kazakidis, V. 2008. A process to improve expected value of mining operations. *Mining Technology* 117(2): 65–70.

Chambers, R.-D. 2007. Tackling uncertainty in airport design: A real options approach. SM thesis, MIT, Technology and Policy Program, August. http://ardent.mit.edu/real_options/Common_course_materials/papers.html.

de Neufville, R. 2006. Analysis methodology for the design of complex systems in uncertain environment: Application to mining industry. Working paper for the Compania Nacional del Cobre (Codelco), Santiago, Chile.

de Neufville, R., Hodota, K., Sussman, J., and Scholtes, S. 2008a. Using real options to increase the value of intelligent transportation systems. *Transportation Research Record,* http://dx.doi.org/10.3141/2086-05. http://ardent.mit.edu.

de Neufville, R., Lee, Y. S., and Scholtes, S. 2008b. Using flexibility to improve value-for-money in infrastructure investments. Symposium on redefining healthcare Infrastructure: Integrating services, technologies and the built environment, Tanaka Business School, Imperial College, London. http://ardent.mit.edu/real_options/Real_opts_papers/Flexibility%20in%20PFI%20projects%20270308.pdf.

de Neufville, R., and Marks, D., eds. 1974. *Systems planning and design: Case studies in modeling, optimization, and evaluation.* Englewood Cliffs, NJ: Prentice Hall. http://ardent.mit.edu.

de Neufville, R., Scholtes, S., and Wang, T. 2006. Valuing real options by spreadsheet: Parking garage case example. *ASCE Journal of Infrastructure Systems* 12(2): 107–111. http://ardent.mit.edu/real_options/Common_course_materials/papers.html.

de Weck, O., de Neufville, R., and Chaize, M. 2004. Staged deployment of communications satellite constellations in low earth orbit. *Journal of Aerospace Computing, Information, and Communication* 1(3): 119–136. http://strategic.mit.edu/PDF_archive/2%20Refereed%20Journal/2_2_JACIC_stageddeploy.pdf.

de Weck, O., Eckert, C., and Clarkson, J. 2007. A classification of uncertainty for early product and system design. ICED-2007–1999, 16th International Conference on Engineering Design, Paris, France, August, pp. 28–31.

ExxonMobil. 2008. *2007 summary annual report.* http://www.exxonmobil.com/corporate/files/news_pub_sar_2007.pdf.

Flyvjberg, B., Buzelius, N., and Rothengatter, W. 2003. Megaprojects and risk: An anatomy of ambition. Cambridge: Cambridge University Press.

Flyvjberg, B., Holm, M., and Buhl, S. 2005. How (in)accurate are demand forecasts in public works projects? The case of transportation. *Journal of American Planning Association* 71(2): 131–146.

Giffin M., de Weck, O., Bounova, G., Keller, R., Eckert, C., and Clarkson, J. In press. Change propagation analysis in complex technical systems. DETC 2007–34652, Design Engineering Technical Conference, Las Vegas, NE. *Journal of Mechanical Design,* MD-07-1230, December.

Hassan, R., and de Neufville, R. 2006. Design of engineering systems under uncertainty via real options and heuristic optimization. Real Options Conference, New York. http://ardent.mit.edu/real_options/Real_opts_papers/Hassan_deN_ROG_submitted.pdf.

Jacoby, H., and Loucks, D. 1972. The combined use of optimization and simulation models in river basin planning. *Water Resources Research* 8(6): 1401–1414. (Reprinted in de Neufville and Marks 1974.)

Lee, Y. S. 2007. Flexible design in public private partnerships: A PFI case study in the National Health Service. Master's thesis, Judge Business School, University of Cambridge. http://ardent.mit.edu/real_options/Real_opts_papers/Yun%27s%20Final%20Dissertation.pdf.

The Peninsula. 2007. RasGas highlights "design one, build many" strategy. http://www.middleeastelectricity.com/upl_images/news/RasGas%20highlights%20design%20one%20build%20many%20strategy.pdf.

Petkova, B. 2007. Strategic planning for rail system design: An application for Portuguese high-speed rail. Master's thesis, Industrial Engineering and Management, University of Groningen, the Netherlands. http://ardent.mit.edu/real_options/Real_opts_papers/petkova%20thesis%20final%20version.pdf.

Shell Technology and Innovation. 2008. Deep water technology. http://www.shell.com/home/content/technology-en/developing_and_producing/deep_water/deep_water_production_12012007.html.

Silver, M., and de Weck, O. 2007. Time-expanded decision networks: A framework for designing evolvable complex systems. *Systems Engineering* 10(2): 167–186.

Suh, E., de Weck, O., and Chang, D. 2007. Flexible product platforms: Framework and case study. *Research in Engineering Design* 18(2): 67–89.

Wang, T. 2005. Real options "in" projects and system design: Identification of options and solutions for path dependency. PhD dissertation, MIT, Engineering Systems Division. http://ardent.mit.edu/real_options/Real_options_papers/Tao%20Dissertation.pdf.

Wang, T., and de Neufville, R. 2006. Identification of real options "in" projects. 16th Annual Symposium, International Council on Systems Engineering (INCOSE), Orlando, FL. http://ardent.mit.edu/real_options/Real_options_papers/Identification%20of%20Option%20in%20Projects%20INCOSE.pdf.

Wikipedia. 2008. Jensen's inequality. http://en.wikipedia.org/wiki/Jensen's_inequality.

9 Real Options in Underground Mining Systems Planning and Design

Vassilios Kazakidis
Laurentian University

Zachary Mayer
Laurentian University

CONTENTS

Throughout their mining life cycle, large multifaceted capital projects in the mineral resource industry are often associated with diverse sources of performance uncertainty. The introduction of flexibility into the planning process is required to counter the downturns and provide the ability to exploit the upturns that can develop over the life of a mining production system.

The quantification of flexibility and the calculation of the option value can provide significant input to the decision-making process. The operating delays in mine production systems are analyzed through process simulation and related to the volatility of parameters controlling the value of a flexible option. The methods to assess the risk associated with a particular mining process and the flexible alternatives considered are discussed, and a methodology is applied to case studies from underground mines in Canada.

The assessment of operating flexibility through the application of indices based on production or economic parameters is also discussed in this chapter.

9.1   FLEXIBILITY IN MINING SYSTEMS

An underground mine is a production system that consists of several excavations/subsystems (Figure 9.1). These subsystems relate to primary development, secondary development, and ore extraction. Underground mine design typically follows an iterative process using economic and risk analyses where potential mining methods are considered, optimum production capacities determined, and indicative cutoff ore grades derived. Primary infrastructure options are then established and the analysis is taken through to global extraction sequencing. Unlike other industrial systems, underground mines are designed, in most cases, before operating conditions are fully understood and the impact of potential operating problems is appreciated. In cases where the uncertainty related to a particular parameter cannot be reduced through design measures, it is important that flexible alternatives be built into the mine plan.

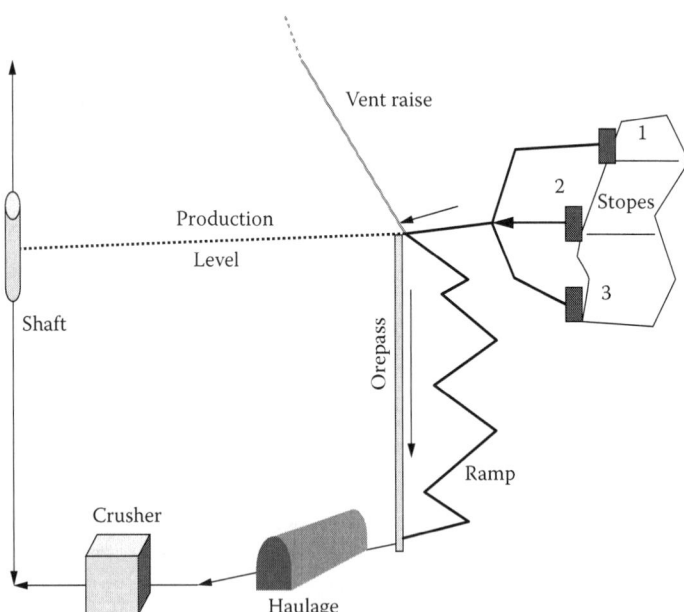

FIGURE 9.1 Schematic layout of a simplified mine production system.

Flexibility is defined as the ability of a system to sustain performance, preserve a particular cost structure, adapt to internal or external changes in operating conditions, or take advantage of new opportunities that develop during a mine's life cycle by modifying operational parameters. Flexibility is an integral part of mine planning and design, and is introduced, often intuitively, into a design and production schedule by the planning team as part of a feasibility study (Kazakidis and Scoble 2001, 2003).

Real options pricing models, developed in the finance sector, have found applications on strategic issues of mineral resource investments (Kajner and Sparks 1992; Sagi et al. 1995; Trigeorgis 1998; Samis and Poulin 1998; Dunbar et al. 1998; Singh and Skibniewski 1991; Kobylanka 2006). Operational options are often termed real options to emphasize that they involve real activities as opposed to purely financial commodities (Luenberger 1998). Until recently, real options analyses could not be applied at a tactical level due to the inability to calculate volatility (Mayer and Kazakidis 2007; Dessureault et al. 2007). However, the introduction of information systems in the mining industry provides the opportunity for real options valuation.

This chapter is structured to first describe the uncertainty present in underground mine production systems and the applicable forms of operating flexibility. After discussing the evaluation of production delays through process simulation, the decision-making process for the analysis of flexible design alternatives is demonstrated using case studies. A methodology for the analysis of flexibility in mining systems is then proposed. Finally, the assessment of operating flexibility through the application of indices is discussed.

9.2   UNCERTAINTY AND FLEXIBILITY IN UNDERGROUND MINING SYSTEMS PROCESS SIMULATION

Uncertainty in a mining project is commonly evaluated with respect to internal (endogenous) and external (exogenous) conditions (Kazakidis and Scoble 2003). The actual return on investment usually differs substantially from that forecasted in the feasibility study, because of the probability of deviations in estimating capital costs, ore reserves, operating costs, mineral revenue, and operating productivity (O'Hara 1982). In mining, uncertainty has been seen as a parameter to calculate risk, while significantly less effort has been made in evaluating opportunities that can be exploited in an operating environment that exhibits uncertainty. Risk management requires an understanding of how these uncertainties can be reduced, the cost of reducing them, and the improvement in project value likely to result from risk reduction. According to Rendu (2002), risk can be reduced in the following ways:

- Acquiring more information
- Choosing flexible engineering solutions
- Improving managerial or technical expertise
- Sharing risk with others

Combining uncertainty and consequence gives a measure of risk. In cases where the risk related to a particular component of a mining system cannot be readily lowered by design improvisations, it is important that a flexible plan be devised to accommodate the factors that can seriously affect production rates or costs, to determine the cost to introduce flexibility, and to determine the potential reward that will follow. Uncertainty has always been a factor in mining and many decisions traditionally rely on the experience and judgment of mine operators (Horsley and Medhurst 2000). Due to the uncertainties inherent in the environment in which mines are planned and developed, it is a given that the introduction of flexibility is essential for the management of risk, the maintenance of low costs, and the preservation of a particular cost structure.

Production simulation (Automod™ and Extend™ are examples of process simulators) is an integral part of mining operations, since it enables the dynamic analysis of complex production systems. Since its introduction to the mining industry in the 1980s, simulation has found application in comprehending the controlling parameters affecting production, evaluating alternatives, predicting the course and results of certain actions, understanding why observed events occur, identifying problem areas before implementation, exploring the effects of modifications, confirming that all variables are known, evaluating ideas and alternatives, identifying inefficiencies, gaining insight and stimulating creative thinking, and communicating the integrity and feasibility of existing plans (Vagenas et al. 2003; Kazakidis and Dessureault 2004).

Incorporating flexibility into the mine plan devised during a feasibility study can help control deviations, and gives management the option to reduce them significantly. Incorporation of flexibility into mine plans will require identification of the applicable areas of a mining system. It was Cavender (2000) who pointed out that the flexibility is introduced to counter an event that would have a negative impact on one or more of the following: production volume, yields (recovery), product quality (grade), production costs, and specified performance. Examples of potential direct implementation of flexibility into an underground mine plan are presented in Figure 9.2 and can be used

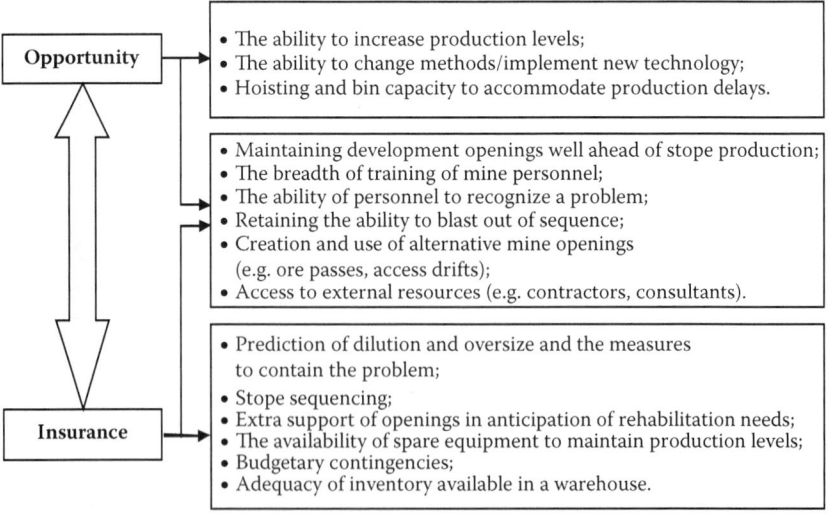

FIGURE 9.2 Examples of direct implementation of flexibility into a mine plan.

as a basis to evaluate alternative scenarios using production simulation (Kazakidis and Scoble 2003; Mayer and Kazakidis 2007).

De Neufville (2003) states that "introducing flexibility into planning and design requires a methodology that entails a deep, almost revolutionary, change in the way designers think about technology, management, and design." In addition to being a valuable asset, flexibility forces designers to do the following:

- Recognize that the value of the project is integrally associated with the fluctuations of the market (or operating conditions) and, thus, they need to be closely in touch with these matters in order to design appropriately.
- Understand that uncertainty is not always a risk to be avoided, but also presents valuable opportunities that can be exploited.
- Adopt a proactive stance toward risk, looking not just to respond to it passively, but also to manage it proactively through the use of real options.
- Introduce far more flexibility into the design of systems than has been the norm.

These concepts find broad application to underground mine design, since a mine is a production system with high operating and market uncertainty.

Production or economic parameters are used in evaluating a decision pertaining to mine design. The economic analysis is directly related to any decision-making process. Real options are important in strategic and financial analysis because traditional valuation tools such as discounted cash flow (DCF) ignore the value of flexibility. If the decision being made involves low uncertainty, or if there is no scope to change course when new information is acquired, then determination of the net present value (NPV) and rate of return (ROR) of a project using traditional DCF techniques would be sufficient. If not, the application of real options can prove to be very important (Copeland and Keenan 1998).

Davies (1997) indicates that a real options analysis requires management estimates of model parameters, including project volatility. In the most basic application of real options pricing, where the value of the underlying project is taken to be stochastic, the project's volatility is one of the most problematic inputs. It is often the case that managers and planning engineers who are comfortable with the NPV method are not accustomed to estimating volatility, because NPV uses expected cash flows and a discount rate based on systematic risk.

Volatility determination that is related to individual project performance will require analysis of operating data records for each process engaged in the project. As pointed out by Copeland and Antikarov (2003), "The volatility of gold is not the same as the volatility of a gold mine." Simulation is an effective tool for estimating volatility; however, it requires managers to provide individual cash flow assumptions and their correlations, rather than provide a single estimate of project risk lumped into one volatility parameter (Cobb and Charnes 2004). According to Mun (2002), the volatility describes the standard deviation of the returns and is a key parameter to estimate in a real options analysis.

There are several methods available to calculate volatility, each of which has its own advantages and shortcomings, based on certain assumptions (Mun 2002). Information technology and data automatically collected and efficiently stored and analyzed, when available, provide a better alternative to calculating volatility than using highly theoretical closed-form equations. Nonetheless, past performance records of a certain mining operation cannot necessarily be applicable in other mines since the conditions (e.g., mining method, geology, stresses, and workforce expertise) can vary significantly.

9.3   MINING PROCESS DELAYS

Assessment of operating flexibility through real options analysis requires the determination of volatility in the controlling system parameters. Production delays can be associated with the variability in performance and the inability to meet preset targets. This section focuses on assessment of mining process delays and determination of volatility through process simulation.

Underground mining often faces uncertainty in production planning and scheduling associated with the performance of the production system components. Production delays can be related to resource planning, equipment performance, and ground conditions. With the high uncertainty associated with mining activities, such delays, due to the adverse performance over time, can have a significant impact on the production schedule and the cost structure of an underground operation. This constitutes a significant difference from industrial production simulations where, unlike with mining, the impact of structural characteristics of a component (e.g., a pylon in an electricity distribution system or the roof of an automotive factory) is not commonly considered as a factor in impairing the overall performance of the production system, unless related to a catastrophe scenario (Kazakidis and Dessureault 2004). The intensity of production delays and economic losses due to ground-related problems can range from insignificant losses up to the loss of a whole year's production (Kazakidis and Scoble 2002).

The causes of delays and their impact on a production system have been the focus of maintenance and reliability analysis in the case of mobile or stationary equipment. An application of maintenance analysis for excavation process performance based on system delay parameters (TBF – TTR) is discussed by Vagenas et al. (2003). A classification of the types of ground-related problems is discussed by Kazakidis and Scoble (2002). Although qualitative analyses have been applied for the prioritization of the controlling parameters in a decision-making process (Kazakidis et al. 2004), it is the quantification of delays that is often required for an economic analysis. The significance of ground conditions and the associated delays in the development of a ramp in a underground mine are shown in Figure 9.3 (Cross 2006). Analysis of such delays can impact a resource planning decision by providing flexibility to counter these ground problems or to improve the quality of the overall system design.

Process simulation is a tool that enables the analysis of complex production system models. It makes possible the evaluation of the impact of system delays and the measures needed to counter them by altering the design/planning parameters of a production system. Although traditionally the focus of process simulation in a mining environment has been on the interaction between crew and equipment (Vagenas 1997; Sturgul and Li 1997; Kelton et al. 2002), the environment in underground mines possesses an additional operating risk factor that can be associated with production delays due to the adverse performance of mine openings over time. The evaluation of the

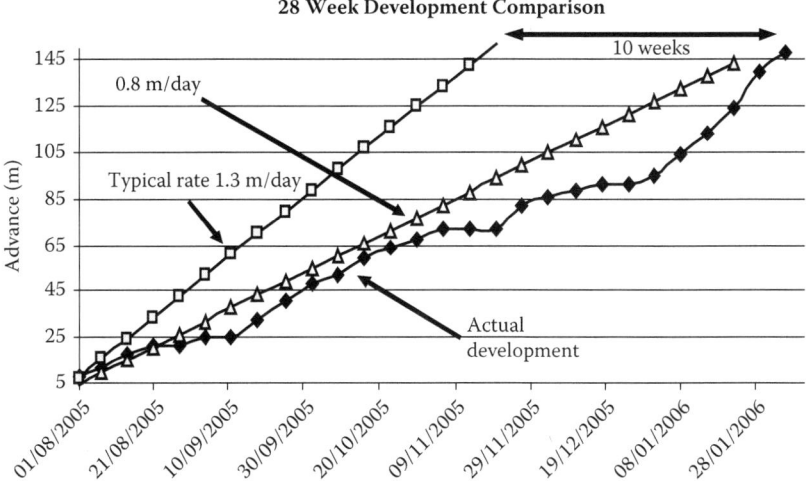

FIGURE 9.3 Planned and actual development rates.

impact of delays related to ground problems in an underground production using a simulation model was examined by Kazakidis and Dessureault (2004). The analysis required the qualification of the frequency and the impact of causes of delays.

Once delay statistics are obtained, a simulation model can indicate the impact of delays in a mine production system (Figure 9.4). The link to economic parameters

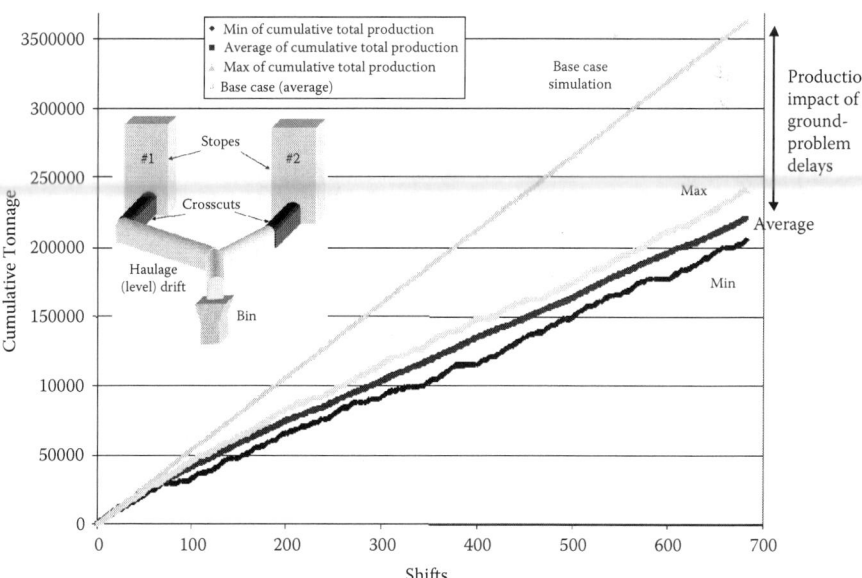

FIGURE 9.4 Results of simulation (20 runs) indicating the impact of ground-related problems on production. Base case includes only operating delays due to crew–equipment interaction.

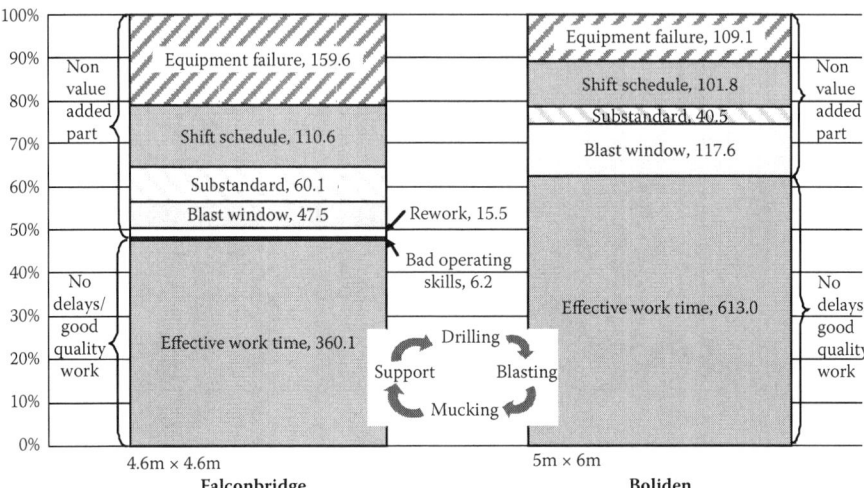

FIGURE 9.5 Time delay impacts for typical development practices with drilling and blasting.

(cost overrun and lost opportunity) is a subsequent step, once the simulation analysis results are obtained. Nonetheless, the impact of flexibility on production parameters can be assessed through simulation, cost-benefit, or sensitivity analyses.

A process simulation analysis using the AutoMod™ software enabled a more detailed analysis of the parameters that impact the development cycle of a mine. The model has been used to evaluate the quality of a particular process in terms of time delays and cost overruns (Kenzap et al. 2007; Kenzap and Kazakid is 2008). The impact of the delays on the development process is highlighted in Figure 9.5 as a result of the simulation analysis. Flexibility components related to available resources (equipment, materials, and labor) can be assessed though the same model in terms of time delays and cost overruns using the statistical parameters that describe the same or alternative processes. The analysis enables design optimization by adding robustness in the system to counter operating problems during its life cycle. The decision-making process over prioritization of alternative design can be performed based on such an analysis.

9.4 DECISION MAKING USING VALUATION OF FLEXIBILITY OPTIONS

Two methods have been considered for the valuation of flexible options based on Monte Carlo simulation: the simplified simulation model analysis, introduced by Winston (2004), and the incorporation of a conditional statement in a cash flow simulation model (Kazakidis 2001; de Neufville et al. 2004).

The *simplified model analysis* uses the Brownian motion to simulate the future value of an underlying asset, which is growing at a risk-free rate and is influenced by its volatility. The method can be used as an approximation, since it considers the same volatility throughout a project. It relates to the entire project parameters (NPV, etc.), making the required inputs simple. The model contains a conditional statement, which triggers an option if, and only if, the value of the project with the option is higher than the value of the project without it. When random sampling is done using Monte Carlo simulation, the range in potential project values, as well as the average value of flexibility, is obtained for the given input parameters.

The *cash flow simulation method* involves the creation of a relatively simple monthly or annualized cash flow spreadsheet with uncertainty adjusted values and a conditional statement (e.g., IF, MAX, and MIN), embedded in the model to make a decision based on the simulated cash flows. Monte Carlo simulation is used to forecast revenue and cost data by using the annualized volatility or by considering the variability of parameters in specific periods in a risk-adjusted model. The result is an NPV based on cash flows generated by the simulation over the life cycle of the project. These models require a trigger to activate the option. The triggers are embedded into a spreadsheet using conditional statements to make the decisions, and can be tailored to suit any specific problem. The value of flexibility is obtained by comparing the results of a spreadsheet with the conditional statement, and one without it; the difference between the two NPV values is the value of flexibility.

9.4.1 CASE STUDY 1: CAPACITY OPTION

An underground mine is evaluating the option of opening a new mining zone. The recommendations from the pre-feasibility study suggest that upgrades be considered to the shaft, skip, hoisting station, and crusher, as well as to construct a new orepass to handle the extra ore produced from the new mining zone (Mayer 2004). The completion of these necessary upgrades will cost the company $68 million over three years. During these three years of upgrading and predevelopment, production in the new zone will be approximately 1,170 tons/day by hauling the ore to the surface using trucks on a preexisting ramp. At the beginning of the third year, when upgrades and construction are completed, the mining zone will produce 5,000 tons/day, generating an NPV of $24.4 million during sixteen years of production.

While upgrading the hoisting infrastructure during the first three years, the company has the ability to increase the hoisting capacity to 6,000 tons/day instead of the currently planned-for 5,000 tons/day. This initial upgrade would in turn give the company the option to pay an extra $6 million for additional orepass and crusher upgrades at Year 3, or later to expand production to 6,000 tons/day, without interrupting the hoisting system operation. The company has two alternatives: the first would be to implement the initial extra shaft upgrades that would give it the flexibility to pay the $6 million for the future production expansion to 6,000 tons/day. The second alternative would simply be to go through with only the originally planned upgrades, and maintain the capacity status quo at 5,000 tons/day. The question is, under what condition would it be beneficial

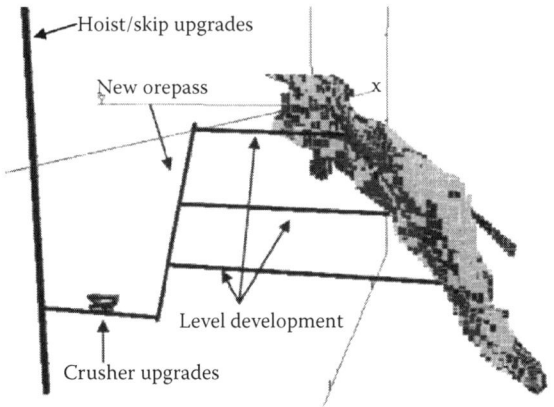

FIGURE 9.6 Layout of the mining zone and required infrastructure upgrades.

to introduce the flexible capacity provided by the first alternative? A layout of the mining zone and necessary upgrades is shown in Figure 9.6.

For this problem, the input parameters that are required for the real options valuation include the project investment cost, the revenues generated by the project, and the risk-free rate. The other variable factors used in the sensitivity analysis are the expansion time, the expansion cost, the expansion factor, and the project volatility (which in this case is controlled mainly by the underlying risky asset, which is the nickel price). The expansion factor is the added percent increase in the project's NPV due to the expansion. Before any calculation can be done, this value must be obtained. For this example, the expansion factor of 9.05 percent is the increased difference in NPV caused by expanding production by 20 percent to 6,000 tons/day up from 5,000 tons/day at Year 3. Because the ore tonnage contained in this zone is fixed, any increase in production will speed up the time frame of the project. In this case, the increase in the project's NPV is due to the reduction in time only, resulting in a modest increase in NPV, given by the expansion factor of 9.05 percent. Figure 9.7 depicts a random Monte Carlo simulation run in Excel (Winston 2004) using the following variable inputs: potential expansion in three years' time, costing $6 million for crusher and orepass expansion.

Time, t =		3 years		Expansion time							
Risk-free rate, r	Current value at time 0	Investment (development) cost at time 0	Value with no option	Volatility (e.g. 20% = 0.2)	Normal (0.1)	Value at time t	Cash flows at time t	Discounted value of cash flows at time 0	Expansion cost, $M	Expansion factor
0.04	92.4	68.0	24.400	0.8	0.64472	97.46	23.62	20.9467091	6	0.0905

	Option to expand = NPV with flexibility − NPV with no option =	$ (3.453)	M

Equations
Value at time t: Current Value *EXP((Risk-free − 0.5*Volatility^2) + Normal(0,1)*Volatility)
Cash Flows at time t (no expansion): MAX((1 + Exp. Factor) *Value at time t − Exp. Cost, Value at time t)
Cash Flows at time t (with expansion): 1.0905*(Value in one year) − 6
Discounted Value of Cash Flows: EXP (-risk free)*Cash Flows at time t − Investment Cost

FIGURE 9.7 Expansion simulation example using Monte Carlo simulation.

A 9.05 percent expansion factor and 80 percent project volatility are also considered. The results generated by this single simulation run show that, in this case, the value of this expansion is worth −$3.453 million, thereby making this option not worthwhile. It is important to note, however, that Figure 9.7 depicts the results of only a single possible path that the stochastic simulation process can take. For this run, the cost of expanding is too high to justify the added value created by the expansion. To get an accurate option value, the model would normally be simulated 20,000 times and the average value used. In this case, the average result for the value of flexibility after 20,000 simulation runs is $0.652 million, thereby making expansion worthwhile.

9.4.2 CASE STUDY 2: SHUTDOWN OPTION

The following is an abandonment option example based on a mine expected to operate for ten years. The risk-free rate is 5 percent, the annual revenue volatility is 40 percent, the annual cost volatility is 10 percent, and at Year 1 the metal price is $4.00/lb. The mine produces 60,000 tons of 1.82 percent grade ore per year. The mine has the option to abandon production at any given year for a cost of $100,000. Figure 9.8 shows the expected cash flows generated by the project for a single simulation run with no option to abandon.

The analysis is based on a methodology proposed by de Neufville et al. (2004). For this model, the metal price and operating costs are simulated for the future periods using the Brownian motion. Alternatively, it is possible to use a mean-reverting model for the metal prices (Copeland and Antikarov 2003; Samis and Davis 2004). In this case, at Year 4 the metal price begins to drop, which causes the operating costs to exceed the revenues generated. The final NPV in this case is −$0.743 million, thereby making the project unprofitable under these simulated conditions. However, the option to abandon production at any year, if metal prices become unfavorable, can improve the project profitability. Figure 9.9 shows the expected cash flows for the exact same project, with the option to abandon embedded in the spreadsheet.

In this case, as soon as the market price drops enough to make the operating costs exceed the revenues, in Year 4, the project is abandoned. From Figure 9.9, the resulting

Year	0.0	1.00	2.00	3.00	4.00	5.00	6.00	7.00	8.00	9.00	10.00
Mineral Pricing Information ($/unit mineral)											
Metal Price ($/lb)	4.000	2.566	3.050	2.833	1.301	1.594	1.216	0.871	0.421	0.490	0.894
Production Statistics											
Production (tonnes mined/period)		60000	60000	60000	60000	60000	60000	60000	60000	60000	60000
Production (tonnes of ore/period)		950	950	950	950	950	950	950	950	950	950
Mineral production (lbs. of ore)		2093800	2093800	2093800	2093800	2093800	2093800	2093800	2093800	2093800	2093800
Operating/Milling/Smelting/Refining cost ($/t)		55.40	58.31	61.69	62.87	59.69	62.11	63.04	63.23	65.53	70.84
Cash Flow Calculation ($ million)											
Operating revenue		5.373	6.386	5.932	2.724	3.338	2.546	1.824	0.881	1.026	1.872
Operating cost		3.162	3.328	3.521	3.588	3.407	3.545	3.598	3.609	3.740	4.043
Risk and Time Adjusted Cash Flow		2.211	3.058	2.411	−0.864	−0.069	−0.999	−1.774	−2.728	−2.714	−2.171
Abandonment cost											0.100
Net Cash Flow		2.211	3.058	2.411	−0.864	−0.069	−0.999	−1.774	−2.728	−2.714	−2.271
NPV Calculation ($ million)											
Time discounted cash flow to time 0		2.103	2.767	2.075	−0.707	−0.054	−0.740	−1.250	−1.828	−1.731	−1.378
NPV	−0.743										

FIGURE 9.8 Cash flow simulation example—without option to abandon.

Year	0.0	1.00	2.00	3.00	4.00	5.00	6.00	7.00	8.00	9.00	10.00
Mineral Pricing Information ($/unit mineral)											
Metal price ($/lb)	4.000	2.566	3.050	2.833	1.301	1.594	1.216	0.871	0.421	0.490	0.894
Production Statistics											
Production (tonnes mined/period)		60000	60000	60000	60000						
Production (tonnes of ore/period)		950	950	950	950						
Mineral production (lbs. of ore)		2093800	2093800	2093800	2093800						
Operating/Milling/Smelting/Refining cost ($/t)		55.40	58.31	61.69	62.87						
Cash Flow Calculation ($ million)											
Operating revenue		5.373	6.386	5.932	2.724						
Operating cost		3.162	3.328	3.521	3.588						
Risk and Time Adjusted Cash Flow		**2.211**	**3.058**	**2.411**	**−0.864**						
Abandonment cost					0.100						
Net Cash Flow		**2.211**	**3.058**	**2.411**	**−0.964**						
RO NPV Calculation ($ million)											
Time discounted cash flow to time 0		2.103	2.767	2.075	−0.789						
RO NPV	6.156										

If (Cost > Revenue) Then shutdown

FIGURE 9.9 Cash flow simulation example—with option to abandon.

NPV with the option to abandon is $6.156 million. Compared to the NPV with no option, this abandonment option adds a significant value to the project. Figure 9.10 depicts graphs representing the operating costs and revenues per period for the project with no option and for the project with the abandonment option incorporated, respectively.

When examining Figure 9.10, it is evident that the project with no option to abandon is operating at a loss from Year 4 and onward. The losses incurred after Year 4 are simply too large to maintain a profitable project. However, as soon as the revenues dip below the costs at Year 4, the project is abandoned. This flexibility to shut down operations saves the company the losses that it would otherwise incur

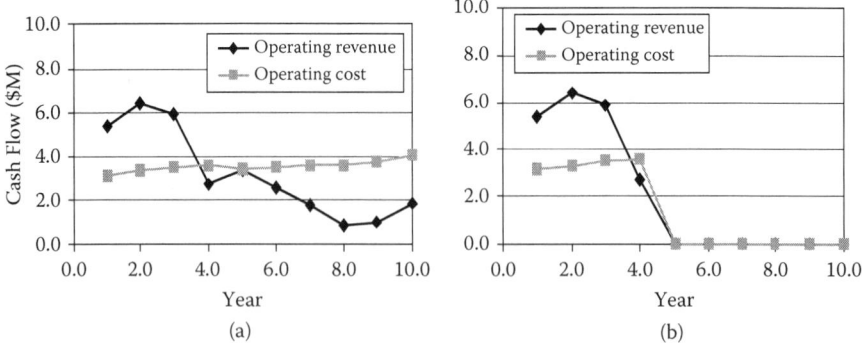

FIGURE 9.10 Revenues and costs per year for (a) project with no option, and (b) project with option to abandon.

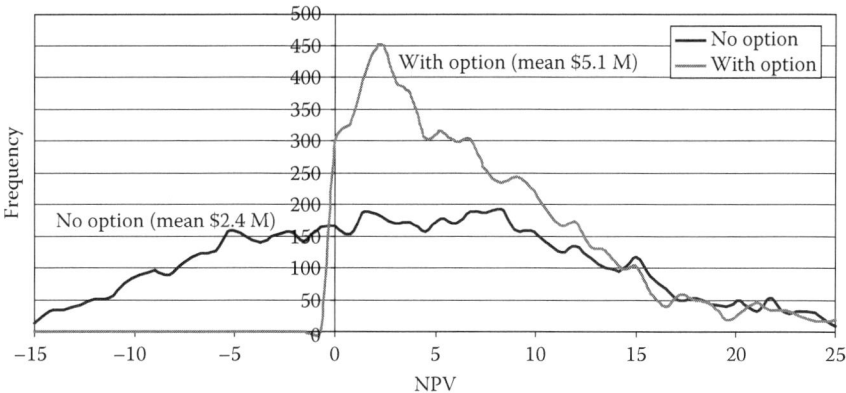

FIGURE 9.11 Distribution of NPVs for no options and with option to abandon.

had it continued to mine. It is clear that this insurance type of flexibility adds a significant value to the project. Monte Carlo simulation was employed to determine the average value of abandonment flexibility for this project after 10,000 simulation runs. Histograms and cumulative frequency charts depicting the range of NPV data were obtained for the two scenarios, as shown in Figure 9.11.

It can be seen that the option to shut down essentially cuts off the low NPV from occurring. As a result, the mean NPV increases from $2.4 million for a project with no flexibility, to $5.1 million for the project that has the flexibility to abandon. Therefore, based on the analysis, it is desirable to introduce this particular option.

9.4.3 CASE STUDY 3: SEQUENCE DECISION OPTION

A valuation for the case of a mine production decision in times of high market uncertainty is considered. The mine must decide which of two possible mining sequences to implement for mining the zone. The first sequence is a full development mining sequence, which lends itself well to the opportunity option of expanding production by taking advantage of rising metal prices. The second possible sequence is a gradual development mining sequence, which is an insurance method that lends itself well to the option of shutting down operations, as a protection against losses caused by a decreasing metal price. The full development sequence requires the ore body to be mined from the bottom up, whereas the gradual development sequence requires it to be mined from the top down. Each of these sequences involves different logistical issues, as well as mining costs unique to the method. A valuation on the two scenarios will help the mine decide which method to implement while recognizing the value of flexibility. The mining zone, shown in Figure 9.12, is a near vertical tabular ore body containing approximately 104,390 tons of 3.05 percent nickel and 1.66 percent copper ore. The ore body, situated approximately 1,000 m deep, has a vertical height of 130 m, a strike length of 150 m, and a width varying between 1 and 3 m.

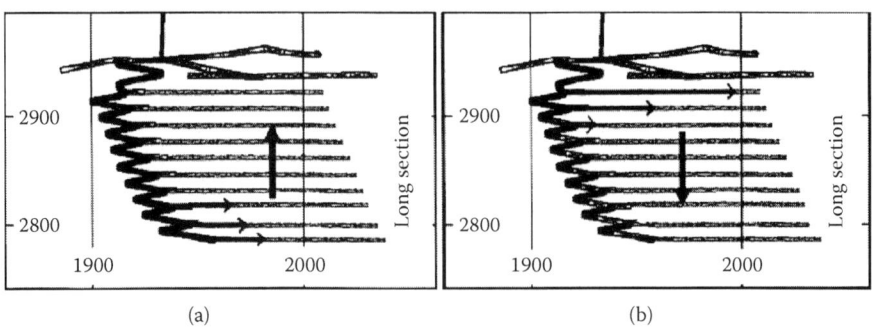

FIGURE 9.12 (a) Full development and (b) gradual development mining sequences.

The mining zone contains ten development levels connected by a spiral ramp (Mayer 2004). A model of the ten-level mining zone was created using the Extend™ discrete event simulation software. The model enables the detailed analysis of a complex production system using actual operating, production, scheduling, and costing data. The simulation model allowed for the outputting of project revenues, volatilities, and time frames for a variety of simulated scenarios.

9.4.3.1 Analysis Sequence 1: Valuation for Full Development Sequence (Opportunity Option)

The expansion scenario involves a decision made by mine management during the construction of the ramp, to increase the number of mining crews from the initial one crew to a three-crew operation. This decision to expand would be made on the basis of the market volatility, as in an upward trend in metal price, which would give the company a beneficial reason to expand by taking advantage of the market upturn in order to maximize profits. A sensitivity analysis was conducted on this model to determine the effects that different project volatilities and expansion costs had on the overall project value. It was found that with a project volatility of 20 percent, expansion is not worthwhile for an expansion cost above $0.8 million. In the case of a project with 40 percent volatility, the option to expand is worthwhile as long as the expansion cost is under $1.2 million. Figure 9.13 represents the trend in project value, as it is affected by varying project volatilities with an expansion cost of $1 million.

Once again, with increasing project volatility, the expansion option adds more value to the project. In this case, with a volatility of 140 percent, the project value can go from $1.81 million by maintaining the status quo, to $2.2 million by exercising the option to expand, as soon as the ramp is complete.

9.4.3.2 Analysis Sequence 2: Valuation for Gradual Development Sequence (Insurance Option)

The shutdown scenario involves a decision made by mine management with the completion of mining at each level to shut down mining indefinitely if conditions call for it. This decision would be made due to a downward trend in the metal price, which

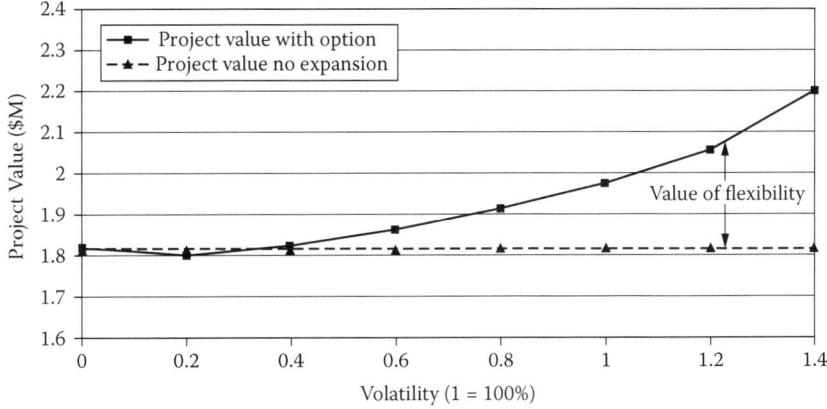

FIGURE 9.13 Volatility sensitivity on project value for borderline expansion scenario.

makes the project unfeasible to continue. The minimum commitment of the gradual development scenario acts as an "insurance" to the project. The value of shutting down varies depending on how far along into the project the company decides to shut down. It can be seen in Figure 9.14 that the value of flexibility gradually increases with time, then eventually peaks and tapers down at the end of the project.

The reason for this trend is that the value of flexibility increases with time, since uncertainty increases with time. However, as mining progresses, the ore body is being gradually depleted, which is what causes the tapering effect of the value of flexibility. It should also be noted that these are all individual European option values, and the mine would have to exercise the option to shut down at only one of these times. Figure 9.15 presents the results obtained from the valuation from both sequences for decision-making purposes.

Determining which of the two development scenarios to introduce in the mine can also prove to be a challenge for the company. Would the mine rather have the option

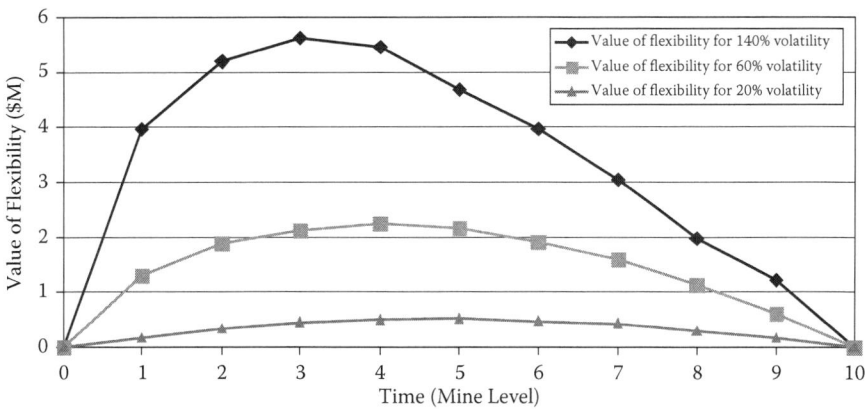

FIGURE 9.14 Value of flexibility per level for various project volatilities.

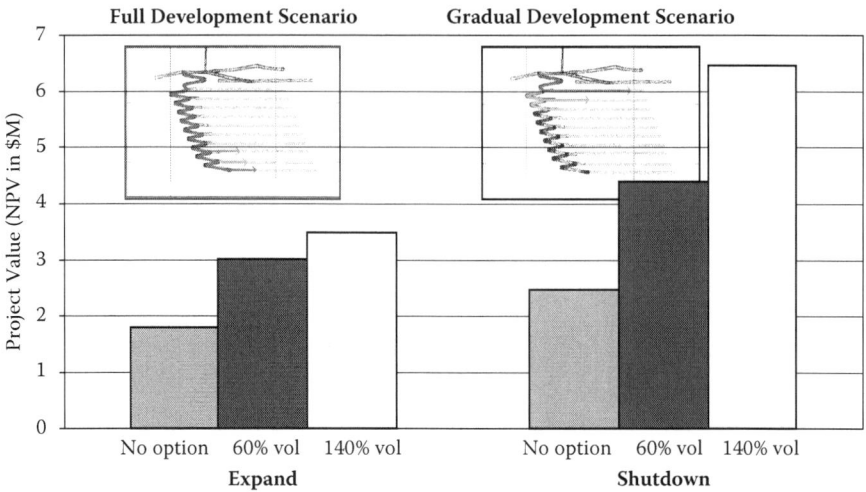

FIGURE 9.15 Comparison of expansion and shutdown effect on project value.

of taking advantage of new opportunities as they arise, or would it rather have the ability to safeguard against a potential loss? It can be seen that the gradual development scenario yields a higher project value then the full development scenario. This shows that for the case study considered, the value of protecting an investment from loss is worth more than the value of trying to increase the investment by expanding operations.

9.4.4 CASE STUDY 4: MINING METHOD SELECTION

In the example shown in Figure 9.16, two alternative mining sequences were considered during the planning of the mining extension of a nickel ore body (Kazakidis and Scoble 2003).

A production schedule for each of the two sequences indicates that the pillarless sequence has a slower start-up than that with pillars. Once the potential production delays and additional operating costs due to ground-related problems are considered and quantified, then it becomes evident that this alone can change the decision over which sequence would be preferable. As an example of the introduction of flexibility into a mine plan, the availability of a rehabilitation crew (the concept is applied to certain mine operations) is evaluated for the sequence with pillars. The cumulative cash flows are shown in Figure 9.17.

It is found that this option improves the overall NPV of the project and, therefore, that it is desirable to introduce this form of flexibility. The results of such an analysis depend on the presumed frequency of ground-related problems and the costs to accommodate the needs for rehabilitation and rework. Should either the intensity of these problems or the cost to introduce measures to minimize the impact on production increase, then the analysis may result in a different conclusion. A sensitivity analysis for the sequence with pillars, with respect to the intensity of ground-related problems, is shown in Figure 9.18.

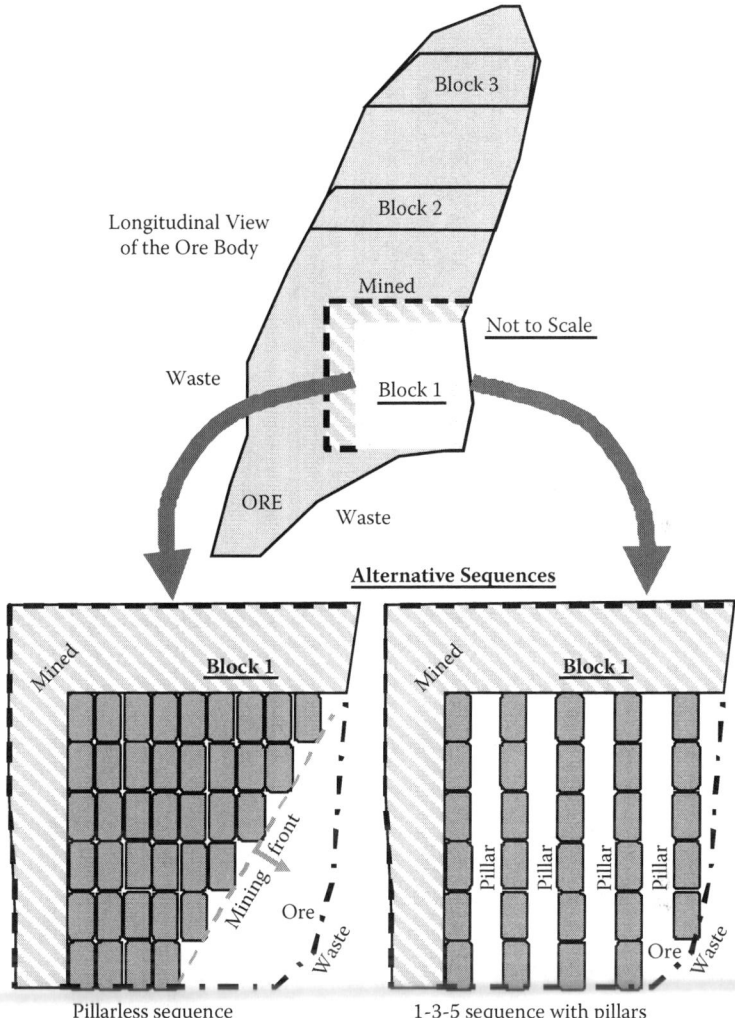

FIGURE 9.16 Consideration of two mining sequences.

It can be seen that, once the annual impact of ground-related problems on lost production exceeds 17,000 tons, the option obtains a positive value, and it becomes worthwhile to include it in the project.

9.5   METHODOLOGY

A methodology that incorporates flexibility into the mine-planning process to better manage the effects of risks is proposed. A flow sheet of the proposed methodology is presented in Figure 9.19 (Mayer and Kazakidis 2007). The four main steps in the proposed methodology are as follows:

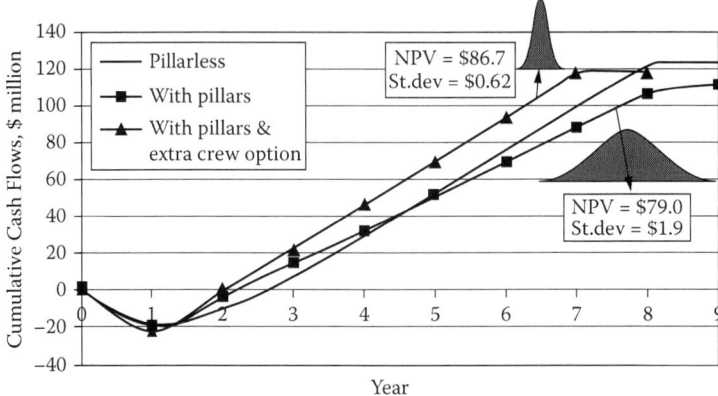

FIGURE 9.17 Cumulative cash flows for the two alternative sequences with ground-related problems using Monte Carlo simulation.

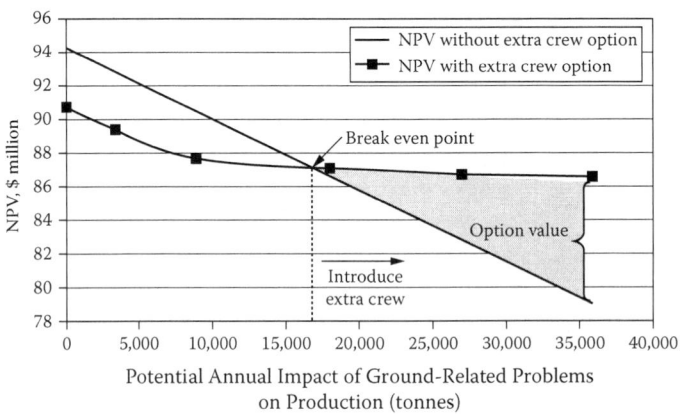

FIGURE 9.18 Sensitivity analysis for the sequence with pillars with respect to the intensity of ground-related problems.

Risk identification: identify potential uncertainties that are relevant to the project in question. A typical risk assessment methodology consists of a search for the answers to the following questions: what can happen, how likely is it that it can happen, and what are the consequences if it does happen?

Process simulation: create a simulation model to assess both production as well as economic data. Such a model can show the direct impacts of various risks on revenues, costs, and other variables. It is at this point that the user must foresee and examine possible problems and consequences that can occur in the mine due to risk factors. Determine how the potential risk reduction methods affect the overall project value or costs.

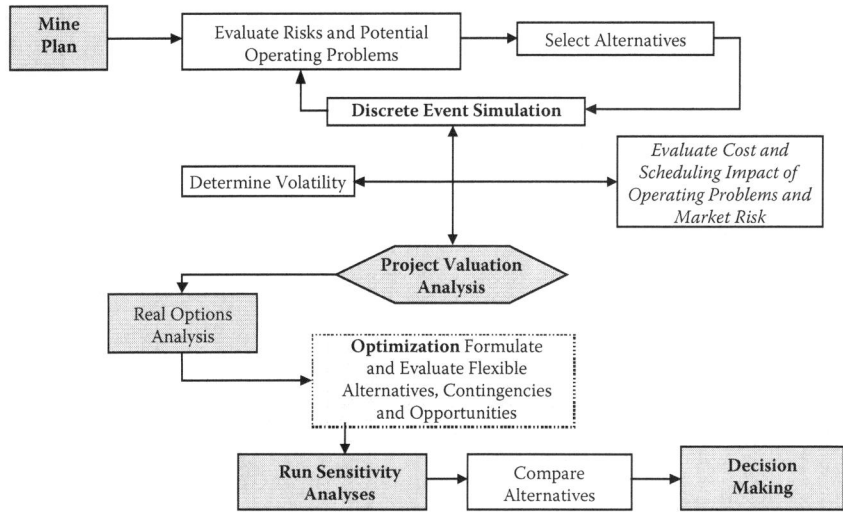

FIGURE 9.19 Proposed methodology flow sheet.

Formulation of flexible alternatives: formulate possible flexible alternatives that can be used to counteract the negative effects of the risks or to take advantage of new opportunities from uncertainties. It is also important to determine under what operating or financial conditions the flexible alternative will be triggered, as well as how much it might cost to do so.

Real options analysis: for the formulated flexible alternatives, use real options valuations to determine the value of incorporating flexibility. Sensitivity of "what if?" analyses can also be conducted to identify the controlling parameters.

9.6 FLEXIBILITY INDEX

The assessment of operating flexibility in the decision making pertaining to mine design is part of the optimization of a plan or part of the evaluation of discrete alternatives. Approaches for developing a flexibility index for mining operations have been discussed by Kazakidis (2001), Woodhall (2002), Kazakidis and Scoble (2003), and Musingwini et al. (2006). It has been stated that despite the fact that in an environment of high uncertainty, such as the one of mining operations, there is need for the introduction of flexibility; this remains a concept difficult to define and analyze. Despite the frequent intuitive introduction of flexibility in mine plans, schedule, and selection of resources, there is no systematic approach in use by the mining industry to measure the flexibility present in the form of an index. When economic parameters are available enabling a real options valuation, an index to be used as a metric for measuring flexibility can be defined in Equation 9.1 (Kazakidis 2001):

$$\text{Flexibility Index, } F\,(\%) = \frac{\text{Option Value, OV}}{\text{NPV passive}} \times 100, \quad \text{OV} > 0 \qquad (9.1)$$

A flexibility index value of 10 percent would indicate that the introduced flexible alternative would improve the NPV of the base case of a project (passive) NPV by 10 percent. This index can be incorporated for measuring a particular type of operating flexibility. It does not in itself, however, relate to the cost of introducing the flexibility.

A flexible alternative is often associated with capital and/or operating costs that have to incur in order for the particular alternative to be active throughout the project. These costs are additional capital outlays, and will occur whether or not the operator exercises the flexible option. This "premium" includes the up-front capital outlays, as well as the additional outlays that may have to occur during the operating stage of a mine to maintain, but not necessarily to exercise, the option, discounted at time zero.

A comparison of the size of this capital cost outlay with the flexibility index can provide a means to examine which of the alternatives are most attractive and would be most valuable to introduce as part of the mine plan optimization. Four flexible alternatives are examined in Figure 9.20.

Alternative A1 (e.g., a second crusher in addition to an existing high-performance crusher) is characterized by a *relatively* low flexibility index value and a relatively high cost to implement it. Implementation of alternative A2 (e.g., an increase in hoisting capacity at a later stage of a mine's life) will have a high impact on the value of the project, but will also have a high implementation cost. Alternative A3 (e.g., an additional vent raise) has both a low cost and a low impact on the value of the project. Finally, alternative A4 (e.g., a second unlined orepass system in a mine) has the highest impact of the four, while its implementation cost is the lowest, and should be the most preferable one.

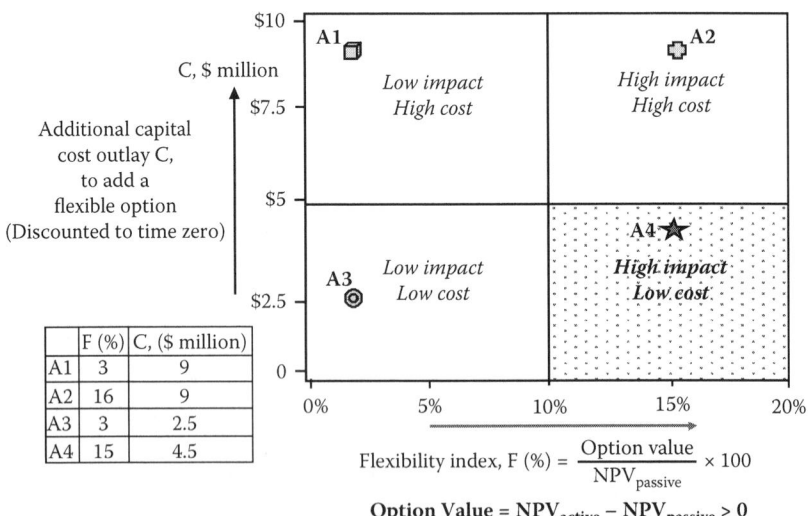

FIGURE 9.20 Classification of flexible options in a mining project using the flexibility index.

The impact that a particular flexible alternative will have on the overall NPV of a project will be a function of the particular characteristics of the mining system and the anticipated operating risk. Costly alternatives such as the lining of an orepass or the increase of the capacity of a hoisting system may prove to be valuable when significant production delays are anticipated related to the performance of particular mining subsystems. In cases of low operating risk, the same alternatives may be found to have only a significantly smaller impact.

Overall, the impact of each design alternative in a flexible mining system would need to be evaluated separately in order to determine its overall impact throughout the life cycle of the production system. This will enable the classification of the various alternatives in the manner shown in Figure 9.20, which should provide a key input to the decision maker for budget allocation and prioritization of flexible alternatives.

Economic data are not always used as a basis to make decisions in an underground mine. Other parameters can be related to production or equipment performance. Woodhall (2002) and Musingwini et al. (2006) propose the evaluation for flexibility in mine layouts based on the availability of production faces (stopes). They define a flexibility index (FI) as

$$FI = \frac{\text{Available fully equiped stopes} + \text{stopes already in production}}{\text{Production stopes required to meet planned production rate}}$$

If FI > 1, then the operation is flexible with respect to the particular parameter (e.g., scheduled stopes). This approach demonstrates the application of a flexibility index to production parameters and can be incorporated into the decision making of an operation. The cost for maintaining a stope open and the likelihood of its need to counter the operating risk in a particular operation should be considered in an economic risk analysis that incorporates these two parameters. Figure 9.21 indicates the decrease in NPV with the increase in ore availability through the introduction of flexibility in a mine schedule (Musingwini et al. 2006).

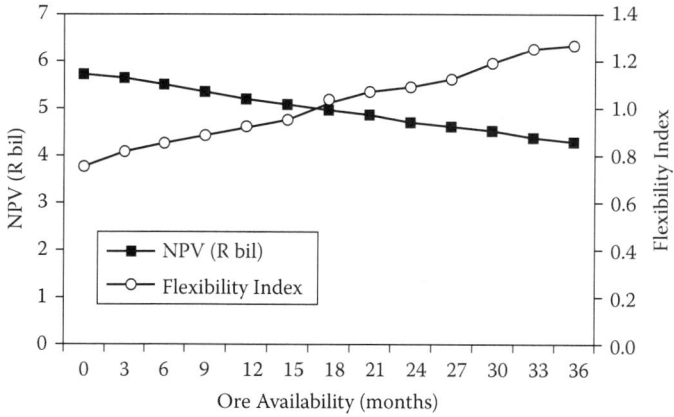

FIGURE 9.21 Variation of NPV and Flexibility Index with ore availability.

9.7   CONCLUSION

Mining systems contain high uncertainty requiring the incorporation of flexible alternatives as part of the planning and design of an underground mine operation. Although flexibility has been introduced in the past intuitively in many aspects of a design, it is the quantification of options under consideration that can impact significantly the decision-making process. This needs the determination of the risk profile and the anticipated system delays, which is usually challenged by insufficient records in a mine operation and their applicability to new operations. Notwithstanding this reality, the demand for more reliable forecasts of production parameters, the need to improve the management of operating and technical risk, and the advancements in information technology applications in the mining industry make feasible the calculation of risk parameters related to operating volatility for the valuation of flexible alternatives. In this process, the introduction of indices to measure the flexibility contained in a particular design is the next step to quantify the robustness of a design against anticipated operating uncertainties.

REFERENCES

Cavender, B. 2000. Rethinking the use of new technology to improve operational performance. *Mining Engineering* 52(12): 61–67.

Cobb, B. R., and Charnes, J. M. 2004. Real options volatility estimation with correlated inputs. *The Economist* 49(2): 119–137.

Copeland, T. E., and Antikarov, V. 2003. *Real options: A practitioner's guide.* New York: Thompson Texere.

Copeland, T. E., and Keenan, P. T. 1998. How much is flexibility worth? *The McKinsey Quarterly* 2:38–49.

Cross, D. 2006. Analysis of delays in the development cycle at a Sudbury mine. Engineering project, Laurentian University, October.

Davies, M. P. 1997. Potential problem analysis: A practical risk assessment technique for the mining industry. *Bulletin of the Canadian Institute of Mining and Metallurgy* 90(1009): 49–52.

De Neufville, R. 2003. Real options: Dealing with uncertainty in systems planning and design. *Integrated Assessment* 4(1): 26–34.

De Neufville, R., Scholtes, S., and Wang, T. 2004. Valuing real options by spreadsheet: A parking garage case example. *ASCE Journal of Infrastructure Systems* 12 (2): 107–111.

Dessureault, S., Kazakidis, V., and Mayer, Z. 2007. Flexibility valuation in operating mine decisions using real options pricing. *International Journal of Risk Assessment and Management* 7(5): 656–674

Dunbar, W. S., Dessureault, S., and Scoble, M. 1998. Analysis of flexible mining systems. *International Journal of Mineral Resources Engineering* 8(2): 165–179.

Horsley, T. P., and Medhurst, T. P. 2000. Quantifying geotechnical risk in the mine planning process. In *Proceedings,* MassMin 2000, AusIMM, Brisbane, Australia, 105–112.

Kajner, L., and Sparks, G. 1992. Quantifying the value of flexibility when conducting stochastic mine investment analysis. *Bulletin of the Canadian Institute of Mining and Metallurgy* 85(964): 68–71.

Kazakidis, V. N. 2001. Operating risk: Planning for flexible mining systems. Ph.D. thesis, University of British Columbia.

Kazakidis, V. N., and Dessureault, S. 2004. Incorporating ground delays in underground mine production system simulation. *SME Transactions* 316:51–57.

Kazakidis, V. N., Mayer, Z., and Scoble, M. 2004. Decision-making using the analytic hierarchy process in mining engineering. mining technology. *Institute of Materials, Minerals and Mining* 113:30–42.

Kazakidis, V. N., and Scoble, M. 2001. Staying flexible in uncertain operating conditions: Integrating contingency into mine planning. 15th Mine Operators' Conference, Sudbury, Canada, February 11–14.

Kazakidis, V., and Scoble, M. 2002. Accounting for the impact of ground-related problems in planning mine production systems. *Mineral Resources Engineering* 11(1): 35–57.

Kazakidis, V., and Scoble, M. 2003. Planning for flexibility in underground mine production systems. *Mining Engineering* 55(8): 33–39.

Kelton, W. D., Sadowski, R. P., and Sadowski, D. A. 2002. *Simulation with Arena,* 2nd ed. Boston: McGraw-Hill.

Kenzap, S. A., and Kazakidis, V. N. 2008. Quality cost assessment in mine tunnelling with drilling and blasting. *International Journal of Mineral Resources Engineering,* 12(3): 161–177.

Kenzap, S. A., Peloquin, L. J., and Kazakidis, V. N. 2007. Use of discrete-event simulation for evaluation of quality in hard rock mine lateral development. *International Journal of Mineral Resources Engineering* 12(1): 49–72.

Kobylanka, M. 2006. The role of real options in the investment decision in mining sector. 2nd International Symposium on Rapid Mine Development, Aachen University, Germany, June 7–8, pp. 549–560.

Luenberger, D. G. 1998. *Investment science.* New York: Oxford University Press.

Mayer, Z. 2004. Real options applications in the valuation of flexible underground mining system design alternatives. MASc thesis, Laurentian University, School of Engineering.

Mayer, Z., and Kazakidis, V. 2007. Decision making in flexible mine production system design using real options. *Journal of Construction Engineering and Management* 133(2): 169–180.

Mun, J. 2002. *Real options analysis: Tools and techniques for valuing strategic investments and decisions.* New York: John Wiley.

Musingwini, C. Minnitt, R. C. A., and Woodhall M. 2006. Technical operating flexibility in the analysis of mine layouts and schedules. International Platinum Conference, Platinum Surges Ahead, the South African Institute of Mining and Metallurgy, Marshalltown, South Africa, 159–164.

O'Hara, T. A. 1982. Analysis of risk in mining projects. *Bulletin of the Canadian Institute of Mining and Metallurgy* 75(843): 84–90.

Rendu, J. M. 2002. Geostatistical simulations for risk assessment and decision making: The mining industry perspective. *International Journal of Surface Mining, Reclamation and Environment* 16(2): 122–133.

Sagi, J. S., Hiob, E. E., and Jones, S. 1995. How option pricing can help the mine manager make decisions. 100th Canadian Institute of Mining and Metallurgy Conference, Montreal.

Samis, M., and Davis, G. 2004. An overview of using real options to value and manage mining projects: Workshop notes. Proceedings 2004 MEMS Pre-Conference Workshop, AGM, Toronto, ON, April 21.

Samis, M., and Poulin, R. 1998. Valuing management flexibility: A basis to compare the standard DCF and MAP valuation frameworks. *Bulletin of the Canadian Institute of Mining and Metallurgy* 91(1019): 69–74.

Singh, A., and Skibniewski, M. J. 1991. Development of flexible production systems for strip mining. *Mining Science and Technology* 13:75–88.

Sturgul, J. R., and Li, Z. 1997. New developments in simulation technology and applications in the minerals industry. *International Journal of Surface Mining, Reclamation and Environment* 11:159–162.

Trigeorgis, L. 1998. *Real options: Managerial flexibility and strategy in resource allocation.* Cambridge, MA: MIT Press.

Vagenas, N., Kazakidis, V., Scoble, M., and Espley, S. 2003. Applying a maintenance methodology for excavation reliability. *International Journal of Mining, Reclamation, and Environment* 17:4–19.

Vagenas, N., Runciman, N., and Clement, R. S. 1997. A methodology for maintenance analysis of mining equipment. *International Journal of Surface Mining and Reclamation* 11: 33–40.

Winston, W. 2004. *Microsoft Excel data analysis and business modeling*. Redmond, WA: Microsoft Press.

Woodhall, M. 2002. Managing the mining life cycle: Mineral resource management in deep level gold mines. MSc project report, University of the Witwatersrand, October.

10 Real Options in Engineering Systems Design

Konstantinos Kalligeros
Massachusetts Institute of Technology

CONTENTS

In this chapter, I examine how the designer can create real options in flexible systems, and how to compare and optimize the flexibility among alternative design solutions. A necessary element in this process is the consistent comparison of the risk exposure achieved by alternative designs, flexible or not. I present a methodology for achieving this, and through an example I demonstrate two key benefits of flexible designs: the potential for significant value creation and the potential to radically change the economic risks in a design by embedding real options.

10.1 ENGINEERING SYSTEMS DESIGN

Engineering systems are always designed, developed, and operated in uncertain economic conditions, for example supply, demand, or prices. These are usually aggregated as a project's "market" or "economic" risk, which is usually crucial to its economic outcome. Moreover, the exposure of a project's success to these economic uncertainties largely derives from technical design decisions. These often determine raw material, product and process specifications, demand limits, and capacity constraints, so that they affect a system's economics through its life cycle. Fixed-point designs are rigid in defining these specifications; flexible designs enable the designer, developer, or operator to actively manage or further evolve the configuration of the system downstream, so as to adapt to changes in the economic environment: they create real options.

The job of the designer of a flexible system is much expanded: it becomes a balancing act between what is technically feasible, the cost of provisioning for downstream flexibility, and the probability that such flexibility may be used. The benefits can be spectacular: we will see that engineering real options in a design not only can increase its value, but also can change its economic exposure to risk. In this chapter, we examine how the designer can create real options in flexible systems, and how to compare or even optimize the flexibility among alternative design solutions. A necessary element in this process is the consistent comparison of the risk exposure achieved by alternative designs, flexible or not.

10.2 FROM FIXED-POINT TO FLEXIBLE DESIGNS

Uncertainty is often perceived as risk that is inherent to a project, regardless of whether it is technical or exogenous (e.g., market related). The system architect's response can range from the trivial (e.g., ignoring risk and designing to an expected value of the uncertainties) to a more sophisticated approach of accounting for risk by imposing sensitivity analysis and "what-if" scenarios on a fixed, predefined design and program development. The latter approach recognizes that risks can devastate the value of a project as well as increase it; it all depends on how the technical specifications (that are fixed in advance) "allow" uncertainties to affect the economic outcome of a project.

A conservative developer will try to eliminate risk altogether by insulating the economic outcome of a system from a particular risk, for example by designing to the most onerous expected value of an uncertain parameter. To illustrate, consider the development of a bridge, where a key design decision is the capacity and a key uncertainty is the vehicle flow. A conservative approach would be to construct the bridge for the lowest anticipated flow (with high probability of materializing), thus insulating the economic viability of the project from this risk. At the same time, this small bridge forgoes economic benefits to the upside. The opposite approach would mean developing the bridge to the maximum potential flow. This would take advantage of all the upside in flow, and at the same time would fully expose the economic outcome of the system to this uncertainty. Assuming higher capacity comes at a higher cost, there is an obvious trade-off between the system's exposure to risk and the economic benefit.

The system designer is regularly confronted with such trade-offs; they are the essence of risk management in engineering systems design (i.e., optimizing a system to reduce the economic downside) and increase the potential upside. In practice, such risk management and optimization are often overlooked (partly because they are tedious). Instead, "point designs" are produced that are optimized to the expected value of the system uncertainties. The economics of a single point design may later be stress-tested to multiple scenarios of uncertainties, but that analysis rarely feeds back to changes in the technical specifications of the system. In contrast, managing risk in engineering design should entail stress testing (under possible future uncertainty scenarios) the economic viability of *multiple* designs in order to select the one whose risk-return profile matches the objectives of the developer. For a bridge, this would mean examining the economic performance of various configurations (e.g., in terms of the number of decks and number of lanes), each under multiple scenarios of vehicle flow. The product of this analysis is no longer a single point design with a corresponding sensitivity analysis; it is multiple designs, each associated with a distribution of future economic outcomes. Cullimore (2001) summarizes this current point-design engineering paradigm, and contrasts it with an approach that comprehensively accounts for uncertainty: "Point designs represent not what an engineer needs to accomplish, but rather what is convenient to solve numerically, assuming inputs are known precisely. Specifically, point design evaluation is merely a sub-process of what an engineer must do to produce a useful and efficient design. Sizing, selecting, and locating components and coping with uncertainties and variations are the real tasks. Point design simulations alone cannot produce effective designs, they can only verify deterministic instances of them."

In this chapter, I build on the "risk management" approach to design, which considers multiple "fixed" design solutions and attempts to compare their risk-return profile in uncertainty scenarios. I extend the concept by including the possibility of expanding, modifying, reducing, or otherwise changing the system in the future, so that it adapts to the uncertain environment. In this approach, I have to consider not only multiple design solutions, but also the technical nature of the transitions between them.

For example, a fixed-point bridge design methodology would consider the expected flow in the future and optimize the design for that expected flow. A risk management approach would require the designer to consider the economic performance of multiple configurations of lanes and decks for the bridge, and later decide on the one with the most desirable distribution of economic benefit, for example a small bridge to mitigate all variability or a large bridge to gain exposure to high potential flow. A flexible bridge design could involve a smaller bridge equipped with such structural enhancements that can carry an additional level in the future, if flow justifies it.

The context and complexity of decisions may vary, but the notion of a "flexible engineering design" has been applied almost identically to systems ranging from phased deployment of communication satellite constellations under demand uncertainty (de Weck et al. 2003); decisions on component commonality between two aircraft of the same family (Markish and Willcox 2003); building design under rent and space utilization uncertainty (Zhao and Tseng 2003; Kalligeros and de Weck 2004; Geltner et al. 1996; de Neufville et al. 2006); and the development of programs of standardized off-shore oil platforms, where standardization at early stages of the program is traded off against flexibility in latter stages (Kalligeros 2006).

So what kind of flexibility can a "flexible design" generate for its developers and users? Table 10.1 attempts to summarize the most important ones. They can

TABLE 10.1

Flexibility in Engineering Systems Design*

		IN DESIGN	IN PRODUCTION	IN USE
MANAGERIAL FLEXIBILITY — **Robustness**	No system reconfiguration	Uncertainty does not affect design decisions. On a design structure matrix, changes in uncertain parameters are decoupled from design decisions	Uncertainty does not cause violation of design requirements.	Feasible (or even optimal) operation for a range of uncertain parameters. Synonymous with resilience. Examples from chemical process design (Biegler, Grossman & Westenberg 1997)
Operational Flexibility	Costless reversible system reconfiguration	Independent task structure matrices (TSMs) or block-dependent TSMs: design decisions can be made without constraining other decisions. Examples from modular computer architectures (Baldwin & Clark 2000).	Process and volume flexibility, usually at the machine or subsystem level. Example: dual-fuel steam boiler (Kulatilaka 1994) or slide-in adjustable-wheelbase automotive chassis (Suh et al., 2004)	Flexibility to switch between alternative states of a system costlessly amounts to ownership of all operational states. Examples: mobile phones that double as cameras, palmtop computers and MP3 players.
Operational Flexibility	Costly reversible system reconfiguration	Sequential, block-hierarchical or coupled TSMs: upstream design decisions constrain downstream ones.	Product flexibility (Sethi & Sethi 1990, Bengtsson 2001), input/output flexibility. Non-zero switching costs incurred with every change.	Similarly to the case above, reversible switching at a cost is equivalent to ownership of all states, but less valuable. Example: flexible office spaces (Greden & Glicksman 2005).
Strategic Flexibility	Irreversible system reconfiguration	The design and implementation of a solution "locks-in" subsequent decisions irreversibly. Examples: legacy systems such as the QWERTY keyboard.	All other irreversible manufacturing decisions, usually regarded as such for modelling purposes. Examples: placement of offshore platforms, BWB aircraft family design (Willcox & Markish 2002).	Irreversible switches in system use, either because of coupled DSMs or high switching costs. Example: expansion/contraction of buildings and infrastructure (Kalligeros & de Weck 2004).

* Managerial flexibility entails only Operational Flexibility and Strategic Flexibility, and excludes Robustness.

Source: From Kalligeros, K. *Platforms and Real Options in Large-scale Engineering Systems Design.* MIT Thesis. 2006. With permission.

be categorized on the basis of the cost (rows) and scope (columns) of system reconfiguration. On one extreme, a system can respond to changes in exogenous uncertainties without any reconfiguration; such a system is *robust* or *resilient* in its life cycle, and often associated with specifications for the most extreme requirements. Engineering systems can also be designed to allow costless and reversible reconfiguration, enabling operational flexibility. For example, a bridge can be designed with an odd number of lanes and the provision of easily allocating the extra lane to the direction of heavier traffic at each time; such a solution has negligible "switching" costs. The next two rows describe costly (but reversible) and irreversible system reconfiguration in terms of the design, production, and system usage. They correspond to the extra structural enhancements that allow building another level on top of an existing bridge.

10.3   COSTS AND BENEFITS OF A FLEXIBLE DESIGN

It can be argued that flexibility is not just an opportunity for added value or a welcome side effect of a good design; in a competitive environment, it should be a design requirement. In the words of Kogut and Kulatilaka (1994), "a myopic policy does not necessarily fail; it fails insofar as uncertainty represents opportunity in a competitive environment." Kogut and Kulatilaka (1994) note that myopic (i.e., rigid) policies in competitive environments fail only as long as uncertainty represents opportunity. For example, de Weck et al. (2003) point out that much of the financial failure of both communication satellite networks built in the mid-1990s, Iridium and Globalstar, could have been avoided if the systems were designed to be deployed in stages, thereby retaining the programmatic flexibility to change the scale of the project, its configuration, and its data-transmitting capacity before its completion.* In real estate, volatile markets have also given rise to flexible development. Archambeault (2002) reports the rapid decline in value for telecommunications hotels after the dot-com market crash at the turn of the century.† With dot-coms failing, the vacancy rate of these buildings increased steeply and their owners started to look into their reconfiguration options. Those buildings that were designed to cheaply convert to laboratory or office space did so at low costs; others had to be drastically redesigned. Similar cases in real estate can be found in Greden and Glicksman (2004).

---

* Iridium and Globalstar were two similar, competing systems of satellite mobile telephony. During the time between their conception, licensing, design, and deployment, a total of about eight years, GSM networks, a competing terrestrial technology, had come to dominate many of the core markets these systems were targeting. Iridium was particularly exposed to such risk, because the system could not operate until all sixty-six satellites were deployed. Eventually, despite their enormous technical success, both companies filed for bankruptcy protection with losses between $3 and 5 billion each. At this time, the two satellite networks operate significantly under capacity with clients such as the U.S. government, exploiting their technical niche of global coverage that terrestrial cellular networks cannot provide.

† Telecommunications hotels are buildings specially configured to host electronic equipment (servers, storage, etc.), servicing the computational needs of Internet companies. The design requirements for such buildings are very different from those of residential or office construction: ceilings can be low, HVAC requirements are stringent, natural light is avoided, and there is very little need for parking.

But the most spectacular benefits of flexible designs have to do with how they change the effect of uncertainty on the value of a project. Consider again the example of a bridge with a fixed number of levels, optimized for the expected demand. For this bridge, larger (expected) variance in demand implies more risk in the revenue stream and less value a priori. A "flexible" bridge that enables future expansion, on the other hand, actually benefits from larger variance in future demand, because larger variance makes it more likely to expand with additional levels and access the additional revenue stream. A point design will always be adversely exposed to more variance in the design requirements; a flexible design may even benefit from more uncertainty. This initially counterintuitive result is perhaps the greatest benefit of flexible designs in uncertain, competitive markets.

On the other hand, these benefits come at a cost: the flexibility to reconfigure a system will rarely come "for free" at the design stage, or be coincidentally available during the operation of a system without any prior provision. (In those cases, the issue becomes how to identify this flexibility and be aware of opportunities to exploit it.) Usually, flexibility implies extra costs at the system design phase, for example in the form of modularization, interface management (see Baldwin and Clark 2000), decoupling of system functions, and platform design; all result in additional development and (possibly) life cycle costs, increased technical complexity, and increased time to market. These costs are the "price" of the flexibility the system designer buys.

Traditionally, declaring the costs of flexibility as "too high" has been the norm. Nevertheless, these costs are easy to quantify; a little harder is to quantify the benefits are. It is necessary though, in order to assess "flexible" design decisions.

10.4 FROM DESIGN FLEXIBILITY TO REAL OPTIONS

The real options associated with flexible, reconfigurable systems are no different from those identified in other contexts of managerial flexibility. As far as irreversible reconfiguration is concerned, it can involve *options to defer investment,* where the holder of the option has the right, but not the obligation, to delay investing a fixed amount to obtain a project whose value is uncertain (McDonald and Siegel 1986), and delaying produces more information about the operational environment of the project. Reconfiguration can involve the *option to defer investment choice,* that is, which project to develop (e.g., see Décamps et al. 2004). The two options of choosing a system reconfiguration and actually going ahead with it are very relevant for engineering systems: they represent the design and construction decisions that most large-scale developments go through. If uncertainties evolve unfavorably during development, engineers have the *option to abandon;* see Majd and Pindyck (1987) and Carr (1988). These are also the options associated with multiple-stage projects or pilot programs that test market conditions before full-scale release of a product: the option to abandon a full-scale deployment is the same as the option to expand the half-scale pilot project. A relevant model by Myers and Majd (1990) shows how a project can be abandoned and its resources can be utilized differently. The *option to grow* the scale of a project makes sense when the expansion path of growth is a sequence of interrelated projects, where earlier stages enable the options to subsequent stages (Kester 1993).

As far as reversible reconfiguration is concerned, *options to switch inputs or outputs* enable the holder to observe the uncertainties as they evolve and make adjustments to the resources required by a project or the outputs produced (Margrabe 1978). A well-known example is given in Kulatilaka (1993), where the value of a steam boiler that can switch from burning oil to gas and vice versa is found as a function of the volatility and correlation between the prices of oil and gas. Such a boiler encompasses the option to switch production inputs, an interesting application in cases where the product mix suffers from internal competition. *Options to alter operating scale* are relevant for projects that can change the scale of production to match demand (for unique products) or market price (for commodities). Such options can be used for choosing among technologies with different characteristics of variable and fixed production costs; see Triantis and Hodder (1990), Tannous (1996), and Brennan and Schwartz (1985). These options include the flexibility to temporarily shut down entirely, perhaps incurring a running "mothballing" cost in order to retain the option to resume production later. In other words, these are reversible options, just like options to switch inputs or outputs: reducing operating scale temporarily does not preclude resuming to full production in the future.

Most projects represent a combination of the above (i.e., they entail *compound options*). Staged projects involve options to abandon or grow to subsequent development phases; each phase may involve options on production scale and speed of development as well as operational options to mothball development; once operational, projects may entail operational flexibility regarding product mix or choice of inputs.

10.5 APPLYING REAL OPTIONS TO SYSTEMS DESIGN

Encompassing any form of flexibility requires two leaps in engineering design. Firstly, just as in the risk management approach to design, it means that multiple scenarios for the drivers of economic performance must be considered. Additionally, multiple alternative system specifications must be developed, together with technical solutions for switching between these configurations. Secondly, it means that the additional costs of system complexity, modularity, and development effort are weighted against a realistic value of the flexibility obtained.

The real options methodology adds value on two fronts here. Firstly, it can help with formalizing the problem of choosing among technically feasible alternative forms of flexibility. Estimating the value of flexibility and selecting among alternative designs can be stated equivalently as the problem of selecting among options of different price (e.g., initial design or construction cost, added complexity, or operational costs) that are "attached" to systems and enable their reconfiguration (at an additional cost, or "strike" price). Formalizing this trade-off can bring clarity into the design process and result in designs of better value. Figure 10.1 shows the trade-off between the initial costs of developing an expandable design (i.e., the price of the real option) and the cost of expansion (i.e., the "strike" price). Assuming everything else equal, cheap designs that can be easily reconfigured (bottom-left quadrant in Figure 10.1) create the maximum value; conversely, solutions that are costly to reconfigure where such capability comes at a high price in complexity, up-front costs, or operational costs (top-right quadrant) create little flexibility value.

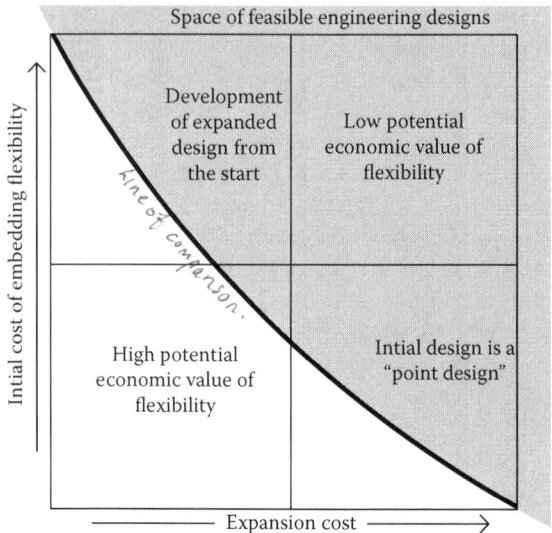

FIGURE 10.1 The trade-off between initial and expansion cost in the space of feasible designs.

Secondly, the real options methodology adds value by enabling the consistent comparison among alternative flexible designs. In Figure 10.1, the choice is not obvious between a cheap-to-construct but expensive-to-reconfigure point design and a cheap-to-reconfigure but expensive-to-construct solution. As long as engineering constraints allow it, there is potentially a lot of value in moving from the bottom-right quadrant of point designs toward the left. Likewise, there can be value in undersizing a system, but enabling the capability to expand the system in the future, moving downward from the top-left quadrant of solutions. The choice depends on engineering constraints, the nature and dynamics of the uncertainties, and the likelihood and potential economic benefit of reconfiguration.

Real options provide a methodology for comparing alternatives along the top-left and bottom-right diagonal of feasible designs, consistently with regard to their risk and economic benefit. The consistency in comparing outcomes enabled with real options analysis can be crucial, because the economic risk of alternatives along this diagonal varies. Flexible systems can be developed to manage this risk, achieving a choice between uncertainty in the economic outcome of the system, and a bottom-line estimate of this outcome. If this choice does not consistently account for risk, then it can be suboptimal.

Despite the relevance of the methodology, however, real options have been slow in reaching engineering managers in practice. One reason is that systems engineering is much more focused on dealing with technical, rather than market, risks;* and

---

* We might distinguish between the two as follows: technical risks involve the possibility that a system does not conform to its technical specifications; market risks involve the possibility that these specifications do not meet the market requirements.

real options are much better suited for market risks. The preference in focus may be because technical risks are easier to quantify and handle, for example by using Six Sigma methodologies. But it is certainly a misplacement of efforts, because market risks by far dominate technical risks for most engineering systems: Iridium is only one of many examples of extraordinary technical successes that have failed due to market-driver uncertainties. Reinertsen (1997) reports on companies that implement Six Sigma methodologies: "the design will fail to conform to its specification at a defect rate of 3.4 parts per million. In contrast, we might look at how successful companies are at meeting market requirements. Most companies would be delighted to have 50 percent of their market share. This corresponds to less than one sigma defect rate." Reinertsen (1997) Reinertsen (1997) reports that Six-Sigma methodologies achieve conformity to technical specifications at defect rates of 3.4 parts per million, whereas companies are usually much less successful in meeting market requirements. He concludes that as technical risks are defined by crisp physical laws, they are a much easier target for engineers than market risks and absorb disproportionately more resources. Another reason for the underutilization of real options in design decisions is the formalistic assumptions and analysis on which the methodology was initially built. While serving academia and economic studies well, this formalism always alienated engineering managers who required a practical, straightforward approach to come to a design decision.

The following text briefly presents one way that real options can be applied to engineering systems design. I present a way to model flexibility as it often appears in engineering systems, and focus particularly on irreversible reconfiguration (i.e., growth, abandonment, and exchange options). I relate these options with a gate-driven process of system development and evolution. Then I briefly present a simulation-based methodology for valuing these options and comparing design alternatives. From a financial standpoint, the methodology has many shortcomings; however, it achieves the consistent comparison of alternative designs from a risk–benefit perspective, which is one of the primary goals of the systems architect–risk manager. As it is simulation based, it additionally fits the framework for risk management followed by many development organizations.

10.5.1 GATEWAY SYSTEM FOR DESIGN AND OPERATIONS

The evolution of many engineering systems is planned as the accomplishment of several milestones or "gates." They usually coincide with fixing specifications and moving on to further stages of development, and in most project design frameworks, these gates are not revisited once passed. In this sense, they are really like the exercise of options, leading to strategic, irreversible design and development decisions. In this section, I show a way to model a gate-stage development process as a series of real options.

The general idea is to model alternative designs or phases of an engineering program as "states" of the program, and the transitions between states as the exercise of options. These transitions occur in two main stages: *design* and *development. Development* can be modeled as the exercise of a call option on the difference in value between the target and departure states, with an exercise price equal to the development cost. This is essentially a timing option, because it involves only the decision of when to begin development and transition to the target state. *Design* involves choosing between alternative target states, and can be modeled as the exercise of an option on the maximum of several underlying assets (a choice option). In the same way that design of a system can lead to its development, exercising a choice option obtains a timing option to build the designed state. In other words, the underlying assets of choice options are the timing options to develop. It is easy to represent these concepts using graphs.

A *state* corresponds to a phase in product development during which a fixed-point design may be in place. It describes a situation of "strategic inaction," a steady-state, business-as-usual situation, where minor, temporary, and reversible interventions to a system's operation are used to maximize its performance; see Vollert (2003, 23). Returning to the bridge example, a state may correspond to any one of the following:

Underwater piles of low bearing capacity in place
Underwater piles of high bearing capacity in place
The operation of a single-level bridge
The operation of a double-deck bridge

The defining characteristic of the states in a model is that the transitions between them are irreversible, so their definition is a modeling decision. States are the building blocks of an options model; graphically, they can be represented with a simple rectangle and a short description of the assets in operation.

The ownership and operation of assets in a state may enable the transition to other states within a time frame, through the expansion, exchange, or reconfiguration of existing assets or the acquisition of new ones. The possible transitions depend on technical constraints and how states are modeled. It may be possible, for example, to enhance low-capacity underwater piles to higher-capacity ones. The transition from low-capacity piles to the operation of a single-level bridge should be possible (perhaps with multiple other states in between). The transition from one state to another involves giving up the value of the "departure" state in exchange for the value of the "target" state. Put in other words, it involves the exercise of an option to exchange one state for another. A simple arrow between states can denote feasible transitions (Figure 10.2).

A nontrivial transition time can be modeled, too; adding time-to-build into the graph would look like Figure 10.3.

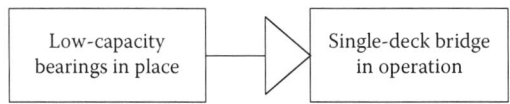

FIGURE 10.2 A timing option to obtain a single-deck bridge in operation, from having low-capacity bearings in place: transition can occur at any time within a time horizon. From Kalligeros, K. *Platforms and Real Options in Large-scale Engineering Systems Design.* MIT Thesis. 2006. With permission.

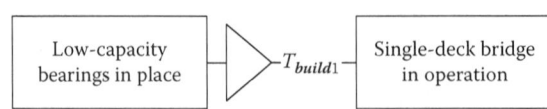

FIGURE 10.3 A timing option to obtain a single-deck bridge in operation, from having low-capacity bearings in place, with time to build T_{BUILD1}. From Kalligeros, K. *Platforms and Real Options in Large-scale Engineering Systems Design.* MIT Thesis. 2006. With permission.

architecture

To model design decisions, we can use "choice options" (i.e., options on the maximum of several underlying assets). In a stage-gate framework, choice options can be used to model design decisions because design involves choosing among alternative engineering solutions, and locking in with these solutions (e.g., by obtaining permits, or securing contracts or other licenses). Returning to the bridge example: the state "underwater piles of high bearing capacity in place" leads to a choice between the operation of a single-layer or a double-layer bridge. To be consistent with how states and development gates are defined, we have to assume that the exercise of a choice option is irreversible; we also assume that whether alive or exercised, choice options do not affect any income generated by existing assets. For example, we assume that the fact that the engineer has a choice for the number of layers of the bridge to be built does not affect the maintenance costs of the high-capacity bearings.

This allows us to "draw" a choice option attached to a state, but separate from it. Figure 10.4 shows the choice option available once the high-capacity bearings are in place. There are two stages in this figure: one involves having high-capacity bearings in place, to designing one of the two bridge solutions and committing to this design; the other involves actually exercising the option to build the (already chosen) number of decks. The graphical representation of the project in Figure 10.4 shows clearly the designer's flexibility.

The entire bridge construction project, with all embedded flexibility, is shown in Figure 10.5. For this example, the bridge design is taken to involve one design decision on the capacity of bearings, which is followed by actual construction and

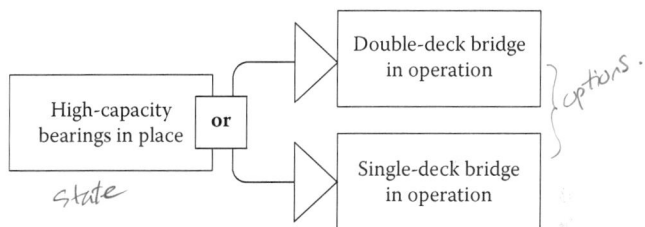

options.

state

FIGURE 10.4 Choice option between a single- and double-deck bridge. From Kalligeros, K. *Platforms and Real Options in Large-scale Engineering Systems Design.* MIT Thesis. 2006. With permission.

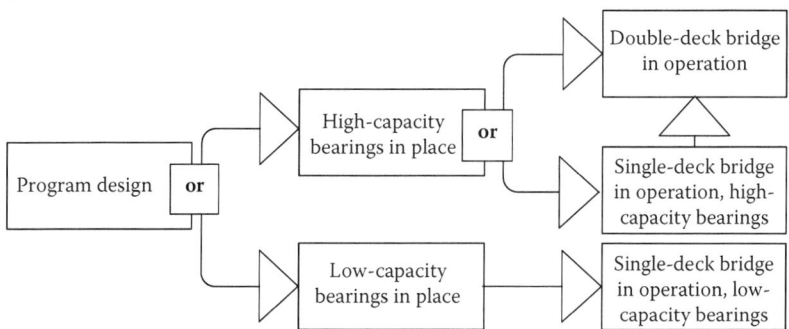

FIGURE 10.5 Design flexibility example: the development of a bridge. From Kalligeros, K. *Platforms and Real Options in Large-scale Engineering Systems Design.* MIT Thesis. 2006. With permission.

binomial tree

leads to that system in place. Low-capacity bearings can only enable the further construction of a single-deck bridge; the state of high-capacity bearings in place encompasses the choice option of further designing either a single- or a double-deck bridge. To simplify, we can assume that a single-layer bridge resting on high-capacity bearings can be expanded by one level without additional design provisions. So far as the same development plan is valid for many alternative designs of bearings and number of lanes for each deck (each one associated with different costs and benefits at each state and construction gate), the value of this plan can be used to compare these alternative designs. In the next section I show how the downside risks, upside opportunities, and embedded flexibility in this design and development plan can be compared on an equal basis in terms of the risk–benefit profile they generate.

10.5.2   RELATIVE VALUATION OF GATES

As explained above, one of the roles of the systems architect is to manage the risk exposure and economic benefits of an engineering system to the exogenous market uncertainties. Management of this risk is necessary for successful designs, and requires the comparison of alternative solutions on the same basis. Applying this to the bridge example, it is evident that a double-deck bridge is much more exposed to variability in demand than a single-deck one: the former will be underutilized for a larger range of realized demand flow, while the former will have a larger probability of operating at capacity. The two designs should be compared on the same basis from a risk–benefit perspective. Extending that notion, a single-deck bridge laid on high-capacity bearings, so that it can be extended by another level, will have a different risk–benefit profile than one whose piles are used to capacity.

This section demonstrates how real options can be used to consistently compare these designs and the options associated with the design (OR) and timing nodes. I first demonstrate why it is appropriate to choose a different discount rate to value each state, OR node, and timing option in a stage and gate representation of a development program, assuming the developer's desired exposure to the underlying risks is the same throughout the development. I then use the bridge example to simulate the uncertainty in development and develop the theory alongside the application, showing how to value the flexibility inherent in a particular design solution.

10.5.3   CHOICE OF DISCOUNT RATE

The intelligent use of a discount rate is at the heart of the methodology described here. Most organizations use a single, flat discount rate for all projects or fixed-point designs. This discount rate appears either directly as such, or as a hurdle rate of return that a project's cash flows must beat. The use of a single number across projects (i.e., designs) of different risk is obviously wrong; but it reveals an important piece of information about the organization's (and system architect's) risk tolerance: if the prescribed rate for discounting economic benefits from a fixed-point design (i.e., operational state) A is R_A, the rate at which the organization can borrow or lend money (i.e., a proxy for the risk-free rate from the point of view of the developer)

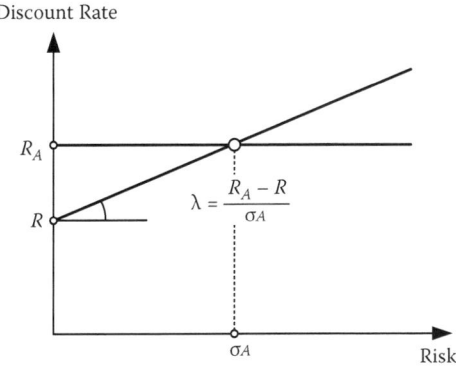

FIGURE 10.6 The implied price of risk given the "accepted discount rate" for design A. From Kalligeros, K. *Platforms and Real Options in Large-scale Engineering Systems Design.* MIT Thesis. 2006. With permission.

is R, and risk is defined as the standard deviation of return σ_A from design A,* then the combination reveals the organization's appetite for risk as $\lambda = (R_A - R_A)/\sigma_A$ (Figure 10.6). This "price of risk" λ should be kept invariant when comparing different design alternatives with different risk profiles; the discount rate used for each one should change to reflect this.†

Consider for example the comparison of the two alternative fixed-point bridge designs, with one and two decks. The standard deviation of returns σ_1 for a single-deck bridge will surely be lower than that for the double-deck design, σ_2, since the former will be fully utilized for a wider range of vehicle flow. Using the same discount rate, for example R_1 for the single-deck bridge, to compare the cash flows from both designs implies using two different prices for risk, λ_1 for the single-deck bridge and λ_2 for the double-deck (Figure 10.7). In this light, it should be much harder to justify the use of two different prices for the same risk than two different discount rates.

Instead, to properly compare the two designs in terms of economic benefit, it is necessary to adjust the discount rate. The double-deck bridge, having larger flow capacity, is more exposed to variability in demand and its cash flow will exhibit a larger standard deviation; therefore, it warrants a higher discount rate. If the same discount rate R_1 is used for this design, it will make it appear economically more appealing than a single-deck bridge. Conversely, suppose R_2 is the benchmark discount rate for a double-deck bridge; then using this rate to evaluate the economics of a single-deck bridge will result in an unfair penalty and make it "look" worse than it actually is. Figure 10.8 shows the two discount rates that are consistent with the amount of risk each solution takes. It does not matter if R_1 is defined first and R_2 second.

---

* The investment return over a time horizon T from a design can be calculated as $R_A = \frac{\text{Value of } A}{\text{Investment cost of } A}$, where the value of the design varies with the market uncertainties, such as demand.

† Technically, this argument is strictly true only when the returns from all alternative designs considered are perfectly correlated. In the case of the bridge, the economic outcome of all alternative designs is assumed to depend on a single source of uncertainty, so the argument holds.

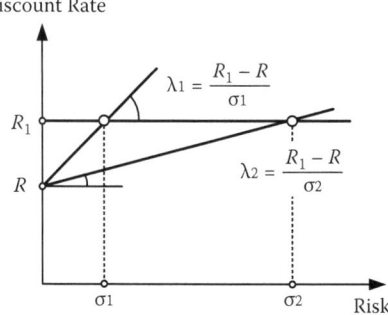

FIGURE 10.7 How the use of the same discount rate for designs 1 and 2 implies a different "risk appetite" for the same uncertainty. From Kalligeros, K. *Platforms and Real Options in Large-scale Engineering Systems Design*. MIT Thesis. 2006. With permission.

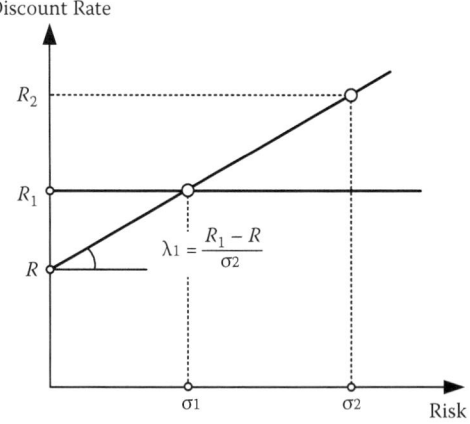

FIGURE 10.8 Assuming the same "risk appetite" for the same uncertainty implies two different discount rates for designs 1 and 2. From Kalligeros, K. *Platforms and Real Options in Large-scale Engineering Systems Design*. MIT Thesis. 2006. With permission.

The concept of a varying discount rate should be applied to all future investment opportunities, not only different states. It should include the valuation of the choice options ("OR" nodes) between different designs, and the valuation of timing options for transition to subsequent states. If there is a single state for which the developer can confidently set the discount rate, then the valuation of all other opportunities can follow consistently. Suppose there is consensus over the risk σ_1 and discount rate R_1 for a single-deck bridge (the "reference state"). The current value V_X of any other investment opportunity X exposed to the same risks can be written as

$$V_X = DF_{X,T} \, E[V_{X,T}(s_T)]$$

where s_T denotes the future value of uncertainties to which the investment is exposed; $V_{X,T}(s_T)$ denotes the value of the investment as of time T, should the uncertain factors

s_T realize; and $DF_{X,T} = (1 + R_X)^{-T}$ is the discount factor corresponding to R_X and T. Denoting the risk-free discount factor $DF_T = (1 + R)^{-T}$, this can be rewritten as

$$V_X = DF_{X,T} \, E[V_{X,T}(s_T)]$$

$$(DF_{X,T}^{-T} + DF_T^{-T} - DF_T^{-T})V_X = E[V_{X,T}(s_T)]$$

$$(DF_{X,T}^{-T} - DF_T^{-T})V_X + DF_T^{-T}V_X = E[V_{X,T}(s_T)]$$

$$V_X = DF_T \Big[E[V_{X,T}(s_T)] - (DF_{X,T}^{-T} - DF_T^{-T})V_X \Big] \tag{10.1}$$

$$V_X = DF_T \, E_{cc}[V_{X,T}(s_T)]$$

What is achieved with the algebra above is that instead of discounting the future expected value of X with the appropriate discount factor $DF_{X,T}$, I use DF_T to discount an adjusted expected value, $E[V_{X,T}(s_T)] - (DF_{X,T}^{-T} - DF_T^{-T})V_X]$. In the end, the same valuation is obtained both ways: either by discounting the real expected value at a rate that reflects the risk inherent in the expectation, or by reducing the future expected value by a certain amount that depends on its risk, and then discounting it at the risk-free rate. The value $E_{ce}[V_{X,T}(s_T)]$ is referred to as the "certainty equivalent value," and the adjustment corresponds to the risk in the expected value of X. This amount is the required risk premium (which is unknown) times the actual value of X at present, which is also unknown. However, if the reference state (e.g., the singe-deck bridge) is exposed to the same risks as X, then their product $(DF_{X,T}^{-T} - DF_T^{-T})V_X$ can be calculated. Holding the price of risk constant between the valuation of a single-deck bridge and any other investment X means

$$\lambda_1 \equiv \frac{DF_{1,T}^{-T} - DF_T^{-T}}{\sigma_1} = \frac{DF_{X,T}^{-T} - DF_T^{-T}}{\sigma_X} \equiv \lambda_X$$

Writing the standard deviation of returns as

$$\sigma_X = StD\left(\frac{V_{X,T}(s_T)}{V_X}\right) \quad \text{and} \quad \sigma_1 = StD\left(\frac{V_{1,T}(s_T)}{V_1}\right)$$

allows us to write the adjustment term as

$$V_X\left(DF_{X,T}^{-T} - DF_T^{-T}\right) = V_1 \frac{StD[V_X(s_T)]}{StD[V_1(s_T)]}\left(DF_{1,T}^{-T} - DF_T^{-T}\right) \tag{10.2}$$

Equation (10.2) follows directly from the capital asset pricing model. Intuitively, it can be interpreted as follows: the amount by which it is appropriate to artificially reduce the value of X due to its risk exposure is proportional to the value of the "reference state" V_1, times the risk premium for V_1, times the ratio of risk in V_X versus V_1.

So the quantity $(DF_{X,T}^{-T} - DF_T^{-T})V_X$ can be estimated based only on what is known about its risk exposure and the valuation of the reference state (V_1) at present time. Substituting this quantity from Equation (10.2) into Equation (10.1) makes it possible to obtain a risk-adjusted valuation for any related investment or design X. The next section describes how simulation can be used to estimate the ratio of standard deviations.

10.5.4 SIMULATION OF UNCERTAINTIES

Equations (10.1) and (10.2) describe how to value consistently any state or option between states, if the standard deviation (i.e., the risk) in the value of these states is known. The valuation is not limited to next-stage options only (i.e., from the present out to a time T). For example, it can be applied to estimate the value of the option to extend the bridge at time $T + DT$, given that the realized demand at T is s_T; in other words, it can be used to value multiperiod options. To achieve this, we need to simulate the uncertain demand over the entire time span $T + DT$. Then, to estimate the standard deviation of future values efficiently, we group (aggregate) the simulation paths together to a number of bins for each period. The methodology follows the "state aggregation scheme" introduced in Rayman and Zwecher (1997), and we apply it directly to the bridge example.

Suppose that the expected demand for the bridge is constant at 10,000 cars per day, and the annual standard deviation in demand growth is 20 percent. Figure 10.9 shows the distribution of demand in 5-, 10-, and 15-year horizons.

To apply the methodology, we follow the steps below:

1. Simulate K paths for the demand flow in a $T = 5$-, 10-, 15- year horizon.
2. At each time step, define m ranges (bins) over the space of realized demand flow (Figure 10.10). A numerically efficient way to do this is to

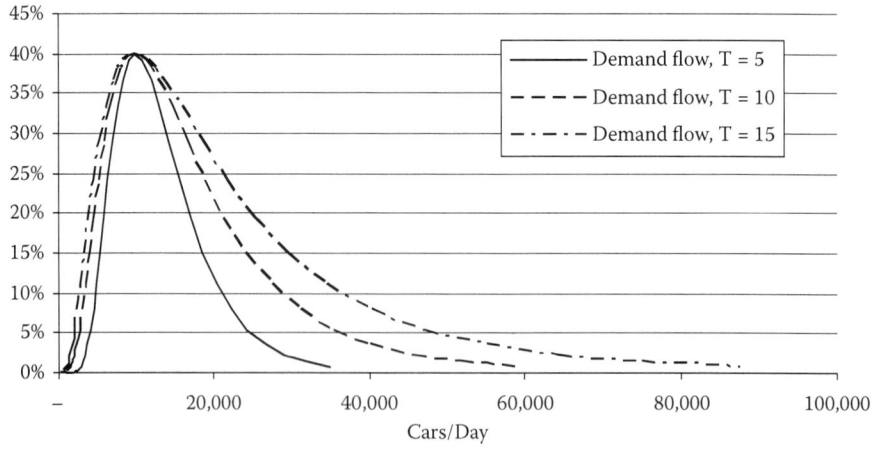

FIGURE 10.9 Distribution of demand flow in 5-, 10-, and 15-year horizons.

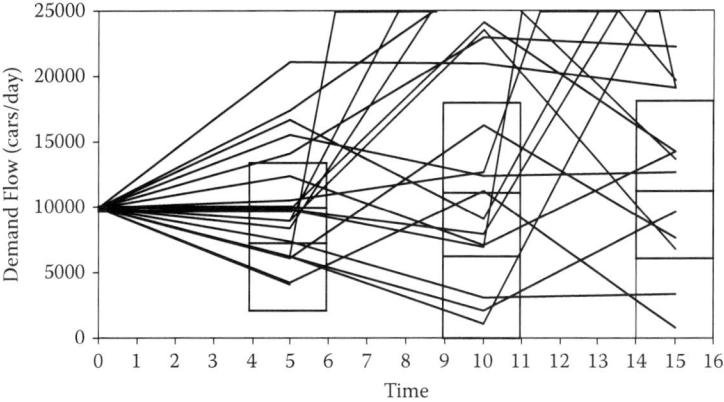

FIGURE 10.10 Schematic of bin definition.

simulate a smaller sample of paths initially and define the bins so that an equal number of paths fall in each bin at any time (Figure 10.10). About 25 paths per bin are usually adequate.

3. For each bin and at each time step, use the full simulation of K paths and take the average value of the paths that fall in that bin. Consider that to be a representative value of the demand flow for that bin, and denote it by $\bar{s}_{t,m}$. For the rest of the analysis, only the representative value of each bin is used.

4. Calculate the transition probability $P(m, n)$ that demand flow is in bin n at time $T + DT$, conditional that demand flow is in bin m at T. This can be estimated as

$$P_T(m, n) = \frac{\text{Number of paths in bin } m \text{ at } T \text{ and in } n \text{ at } T + DT}{\text{Number of paths in bin } m \text{ at } T}$$

Applying the methodology to the bridge example for four bins per period, we obtain the bins shown in Table 10.2 for each time period. The corresponding transition probabilities are shown in Table 10.3. For example, the probability that a path falls in bin 2 at $T = 5$ (i.e., that realized demand flow is between 7,338 and 9,974 at the time), and bin 3 at $T = 10$ is 23 percent.

10.6   VALUATION OF A DESIGN SOLUTION

It is now possible to value the flexible design solution in the uncertain demand flow environment. The structure of flexibility created with a single-deck solution that is expandable to a second level was defined in Figure 10.5. Suppose the capacity of the proposed design is 10,000 vehicles per day for the single-deck bridge and 20,000 for the double-deck one, and that each vehicle generates a revenue of $1.00 (which cannot change). While there are potentially numerous feasible technical solutions with

TABLE 10.2

Bin Definition for Simulated Demand Flow Uncertainty

m	min $S_{T,m}$	max $S_{T,m}$	$\bar{s}_{T,m}$	max $S_{T,m}$ – min $S_{T,m}$
$T = 5$				
1	2104	7328	5671	5223
2	7338	9974	8679	2636
3	9989	13501	11631	3512
4	13502	43511	18687	30009
$T = 10$				
1	1	6260	3614	6260
2	6264	11251	8714	4987
3	11254	18217	14564	6963
4	18241	1170750	33370	1152509
$T = 15$				
1	49	6136	3533	6086
2	6142	11310	8630	5168
3	11320	18381	14500	7061
4	18385	414037	29141	395652

TABLE 10.3

Transition Probabilities between Bins

	Bins, $T = 5$			
$T = 0$	1	2	3	4
	25%	25%	25%	25%

	Bins, $T = 10$			
Bins, $T = 5$	1	2	3	4
1	52%	23%	14%	11%
2	29%	34%	22%	15%
3	15%	30%	32%	24%
4	4%	14%	32%	50%

	Bins, $T = 15$			
$T = 10$	1	2	3	4
1	40%	27%	19%	15%
2	25%	30%	21%	23%
3	19%	24%	30%	26%
4	16%	19%	30%	36%

TABLE 10.4

Design and Construction Costs

Design bearings (high or low capacity)	$ 1,000,000
Design deck (first or second)	$ 1,000,000
Construct high-capacity bearings	$ 10,000,000
Construct low-capacity bearings	$ 7,000,000
Construct single deck	$ 8,000,000
Construct double deck	$ 13,000,000
Construct additional deck on single bridge	$ 5,000,000

different capacity and expansion cost levels, we will examine the one summarized in Table 10.4.

First, we will extract the stakeholders' risk attitude based on the valuation of a reference state, for example the operation of a single-deck bridge resting on low-capacity piles. Assume that a discount rate $R_1 = 6.0\%$ is acceptable for that. Second, we need to decide on the time horizons for which the options are relevant. We have already implicitly decided on this by simulating and aggregating the demand flow uncertainty for 5, 10, and 15 years. Using this information, we will value all the associated options and design choices shown in Figure 10.5, working backward from the final stages of development (i.e., a fully operational single- or double-deck bridge) to the initial design choices. Each option will be valued at the 5-, 10-, and 15-year "decision points."

10.6.1 VALUATION: SINGLE-DECK BRIDGE (REFERENCE DESIGN)

An inaccurate analysis would involve simple annual discounting of expected cash flows, and put the present value of a single-deck bridge (i.e., the reference design) at $45,717,846. This is obviously wrong as it doesn't capture the fact that capacity is capped at 10,000 and it is uncertain. A better analysis should calculate the present value of the bridge for each of the four representative levels for demand flow at $T = 5$, that is, 5671, 8679, 11631, and 18687 vehicles/day. These yield future present values of $25,924,858, $39,680,625, $45,717,846, and $45,717,846, respectively. The last two values are equal because any demand level above 10,000 cannot be accessed and will not generate revenue. Using the transition probabilities in Table 10.3 (i.e., 25 percent to each bin), we calculate the expected present value as

$$V_1 = \$39,260,294$$

and the standard deviation of future values at 20.6 percent. The analysis is repeated identically for each representative $\bar{s}_{T,m}$ at each $T = 5, 10, 15$ and each bin $m = 1 \ldots 4$, to obtain Table 10.5. For example, $38,219,224 is the future value at $T = 5$, provided that realized demand is 5,671.

TABLE 10.5

Valuation of Reference Design

	Bin	$\bar{s}_{T,m}$	
$T = 0$	1	10000	$39,260,294
$T = 5$	1	5671	$38,219,224
	2	8679	$45,209,387
	3	11631	$51,913,748
	4	18687	$58,830,669
$T = 10$	1	3614	$40,473,358
	2	8714	$46,602,256
	3	14564	$49,979,000
	4	33370	$51,318,926
$T = 15$	1	3533	$20,692,879
	2	8630	$50,542,694
	3	14500	$58,568,984
	4	29141	$58,568,984

In the end, we have the value of the reference state as well as all the needed information to adjust the expected values in other states for risk. Assuming a risk-free discount rate $R = 5.0$ percent, Equation (10.2) gives for $T = 1$ (a single period)

$$V_X \left(DF_{X,T}^{-T} - DF_T^{-T} \right) = V_1 \frac{StD[V_X(s_T)]}{StD[V_1(s_T)]} \left(DF_{1,T}^{-T} - DF_T^{-T} \right)$$

$$= 39,260,294 \frac{StD[V_X(s_T)]}{20.6\%} (6.0\% - 5.0\%) \qquad (10.3)$$

$$= \$1,906,670 \, StD[V_X(s_T)]$$

This is interpreted as follows: if the future values of another state or option X have a standard deviation of 100 percent, then the value adjustment to its present value should be $1,906,670, or 4.86 percent of the expected value of the single-deck bridge.

If we consider no flexibility in postponing development or expanding the bridge, the single-deck solution would be valued as a fixed point design. The construction cost would be subtracted from the value of the single-deck asset (i.e., state "1"), and the net present value would enter the investment choice:

$$NPV_1 = V_1 - \text{Construction cost} = \$39,260,294 - 17,000,000 = 22,260,294$$

10.6.2 Valuation: Double-Deck Bridge

Repeating the analysis for the bridge with double the capacity yields values of $42,793,994, $50,458,672, $61,597,768, and $86,960,264 for the four levels of demand at $T = 5$. The values are not capped, since the two-deck bridge capacity is always larger than the representative values of demand at $T = 5$. The standard deviation here is 43.2 percent, the risk adjustment calculated from Equation (10.3) is $888,108, and the present value is $41,425,200. Comparing the two fixed-point designs, the higher present value of the double-deck bridge is offset by the increased construction cost, so that in the end,

$$NPV_{B2} = V_{B2} - \text{Construction cost}_2 = \$41,425,200 - \$25,000,000 = \$16,425,200$$

so the single-deck bridge appears to be a better design.

10.6.3   VALUATION: OPTIONS TO BUILD ON HIGH-CAPACITY BEARINGS

The next step is to value the flexibility available to the operator once high-capacity bearings are installed. We will call this state "HC." In itself, it generates no income, so it has no intrinsic value. It enables the choice, however, between designing a single- or double-deck bridge, which we will denote as "OR2," and the value of the right to choose, V_{OR2}. Each design choice enables the construction of its respective configuration, that is, it triggers the option to construct a double- or a single-deck bridge. These options are denoted "HCB2" and "HCB1," respectively, and we write their present value as V_{HCB2} and V_{HCB1}. The value of the double-deck bridge, V_{B2}, was calculated in Section 10.6.2. The value of the single-deck bridge is more than the value V_1 in Section 10.6.1, because it represents more than just the cash flow from operations: it also enables the construction of a second deck. This is an option, denoted "B1B2" with value V_{B1B2}, leading to state "B2." The sequence of calculations should be intuitive: start valuing the option B1B2; then each of the options to construct B1 and B2, respectively; and finally the option to choose, OR2. V_{OR2} will be the value of having high-capacity bearings installed.

To calculate V_{B1B2}, assume that expansion is instantaneous, so that the payoff of expanding at time T and node m is

$$\text{Value of Expansion} = V_{B2}(s_{T,m}) - V_{B1}(s_{T,m}) - \$5,000,000$$

At each node, the operator can either expand immediately or wait and possibly expand at the next opportunity in five years. The value of waiting is given by Equations (10.2) and (10.3) as

$$W_{B1B2}(s_{T,m}) = DF_5 E_{cc}[V_{B1B2}(s_{T+5,m})]$$

where the discount factor DF_5 is taken to be the risk-free discount factor five years ahead* and the certainty-equivalent expectation is taken over the nodes $m = 1 \ldots 4$ at

---

* The appropriate discount factor to use at T is the one from T to $T + 5$, but if the interest rate yield curve is flat, as it is assumed here for simplicity, then this will be equal to the discount factor from zero to five years.

time $T + 5$. Expansion is optimal at time T and node m only if the value of expansion is greater than the value of waiting, that is,

$$V_{B1B2}(s_{T,m}) = \max\ [V_{B2}(s_{T,m}) - V_{B1}(s_{T,m}) - \$5,\ 000,000;\ W_{B1B2}(s_{T,m})]$$

For $T = 15$, we assume there will be no other opportunity to expand in the future, so the operator will expand instantaneously if it makes sense (i.e., if the value of expansion at the time is larger than zero). Starting from $T = 15$ and working backward, the value of B1B2 at $T = 0$ comes out to

$$V_{B1B2} = \$6,157,707$$

and raises the total value of a single-deck bridge resting on high-capacity bearings to

$$V_{B1} = V_1 + V_{B1B2} = \$45,420,001$$

To obtain a single- or a double-deck bridge, the operator must first build them (incurring a construction cost of $8 million for B1 and $13 million for B2). Valuing each of the two options exactly the same way as above yields

- Option to construct B1 on high-capacity bearings: $V_{HCB1} = \$38,732,748$
- Option to construct B2 on high-capacity bearings: $V_{HCB2} = \$36,972,943$

To obtain either option, the developer must first choose to design either B1 or B2 (i.e., exercise the OR2 option available with the HC state). The formulation is similar: to exercise OR2 obtains the greater of V_{HCB1} and V_{HCB2} minus the design cost ($1 million for both designs). To wait at each period and possibly decide in the future is calculated using Equations (10.2) and (10.3) again. So the waiting value at time T and node m is

$$W_{OR2}(s_{T,m}) = DF_5 E_{cc}[V_{OR2}(s_{T+5,m})]$$

where the discount factor DF_5 is taken to be the risk-free discount factor five years ahead and the certainty-equivalent expectation is taken over the nodes $m = 1 \ldots 4$ at time $T + 5$. The value of OR2 at each node is based on optimal exercise:

$$V_{OR2}(s_{T,m}) = \max\ [V_{HCB1}(s_{T,m}) - 1,\ 000,000;\ V_{HCB2}(s_{T,m}) - 1,000,000;\ W_{OR2}(s_{T,m})]$$

and the value at the current initial node ($T = 0$) comes out $V_{OR2} = \$37,940,123$. Notice that this is larger than both V_{HCB1} and V_{HCB2} (after subtracting design costs), implying that the option to postpone the design choice is valuable. Indeed, the calculation of the option at each node reveals that the optimal action for a hypothetical owner of high-capacity bearings at $T = 0$ is to wait and design B1 if flow is low at $T = 5$, or B2 if it is high (Table 10.6).

TABLE 10.6

Optimal Actions for Option OR2 to Design Either B1 or B2

m	T = 0	T = 5	T = 10	T = 15
1		B1	B1	B1
2		B1	B1	B1
3	Wait	B1	B1	B2
4		B2	B2	B2

10.6.4 VALUATION: OPTIONS TO BUILD ON LOW-CAPACITY BEARINGS

Flexibility for the operator of low-capacity bearings (state LC) is limited to timing the design and construction optimally. Following the same process described above, we obtain for the option to construct a single-deck bridge (nonexpandable) on low-capacity bearings, V_{LC1} = \$38,732,748.

10.6.5 VALUE OF FLEXIBILITY

The initial choice faced by the developer is the capacity of bearings to design and then build. Proceeding exactly as before, we calculate the initial value for these options:

- Option to construct high-capacity bearings: V_{0HC} = \$30,264,813 realized by waiting at $T = 0$
- Option to construct low-capacity bearings: V_{0LC} = \$25,940,727 realized by waiting at $T = 0$
- Choice option between 0LC and 0HC: V_{OR1} = \$29,616,038 realized by waiting at $T = 0$

Since OR1, the initial choice between high- and low-capacity bearings, is the first action in the development plan of the bridge, it is worth examining in further detail the optimal strategy revealed from the analysis for (Table 10.7). If the developer waited (suboptimally) until Year 15, then the best strategy would be to proceed with low-capacity bearings for the two low-demand events. On the other hand, if the developer waited until Year 10, then the optimal choice would be high-capacity bearings for all but the lowest-demand event at the time. This is because there exists a further option downstream the development plan, which is to build a single-deck bridge on high-capacity bearings. At $T = 5$, the optimal strategy is to develop high-capacity bearings, unless the lowest-demand event materializes, in which case the developer should wait further. Finally, at the initial stage, the optimal strategy is to wait. Table 10.7 shows clearly what timing and choice flexibility is created by staging the project.

The value of this flexibility follows directly: the initial choice OR1 is worth V_{OR1} = \$29,616,038. The net present value of developing a single-deck bridge

TABLE 10.7

Optimal Actions for Option OR1 to Design Either HC or LC

m	T = 0	T = 5	T = 10	T = 15
1		Wait	LC	LC
2		HC	HC	LC
3	Wait	HC	HC	HC
4		HC	HC	HC

without any expansion flexibility was calculated above, $NPV_1 = \$22,260,294$. And that of a double-deck bridge was found to be $NPV_{B2} = \$16,425,200$. Adding expansion flexibility to the single-deck bridge (the best point design) increased its value by an impressive 33 percent.

This result depends heavily on how the design and construction costs compare at each development stage and among the two design solutions. That is exactly the point where technical expertise is crucial. Suppose it is possible to plan the construction process so as to reduce the cost of building a double-deck bridge directly on the high-capacity bearings to $10 million. Good project managers may be able to achieve that by cutting the cost of multiple mobilizations that are unavoidable in staged projects. As we would expect, the net present value of a double-deck bridge would be exactly $3 million higher (because the cost is $3 million lower), while the value of the initial choice would rise less (approximately $1 million). Different technologies (i.e., different combinations of initial costs, target capacities, etc.) comprise the design space of Figure 10.1. The engineer's job is to find the optimal feasible solution, considering the risk.

A final note on risk: suppose we increase the level of demand flow uncertainty, σ, from 20 to 30 percent, but take care to maintain the value of risk in the analysis constant at 4.86 percent per 100 percent of standard deviation (to keep the developer's risk appetite constant). By trial and error, we find that this implies a discount rate of 6.37 percent for the single-deck bridge. The higher variability for the same risk appetite would make all fixed-point designs less valuable, and particularly the single-deck bridge because this solution does not access any higher potential revenue that becomes more likely with higher variability. The analysis shows that its net present value drops from $22,260,294 to $17,710,241 (Figure 10.11). By contrast, the net present value of the double-deck bridge stays practically the same: the higher probability of high demand flow is compensated consistently by the higher risk in the higher discount rate. The value of the flexible design drops marginally from $29,626,038 to $28,139,839, that is, 59 percent higher than the value of the single-deck bridge given the same level of uncertainty. The marginal decrease in the value of the flexible design makes sense because the value of the single-deck bridge is lower should that be developed at $T = 15$; however, it is possible that alternative technological solutions can even cause the value of a flexible design to increase as uncertainty increases. The increase in the value difference (59 percent versus 33 percent) is due to the higher

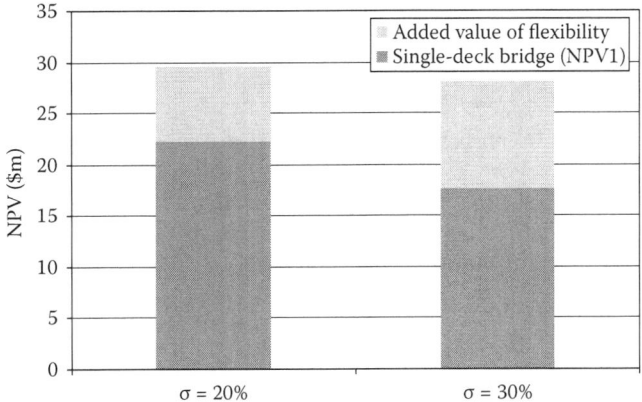

FIGURE 10.11 Embedding flexibility in the single-deck bridge improves the risk exposure of the design.

value of flexibility in the development plan. Evidently, embedding flexibility in the single-deck design has drastically reduced the effect of risk on its value.

10.7 CONCLUSIONS

In this chapter, I have made several important points on "flexibility in engineering systems" that I summarize below.

- Economic risk can be crucial to the economic outcome of any design; therefore, the comparison of alternative designs must account for such risk. One approach is to compare the economic performance of a design under multiple scenarios of the future uncertainties, so that comparison is based on a distribution of economic performance rather than the performance over an expected economic scenario. This amounts to a risk management approach to design, and it is necessary before any discussion of flexibility in design.
- A flexible design is simply one that enables the designer, developer, or operator to actively manage or further evolve the configuration of the system downstream, so as to adapt to changes in the supply, demand, or economic environment. Most forms of such active management can be formulated as real options problems.
- Under a risk management design approach, the job of the designer of a flexible system becomes a balancing act of what is technically feasible, the cost of provisioning for downstream flexibility, and the probability that such flexibility may be used.
- In the suggested framework, designers decide on the flexibility they want to enable (i.e., the real options) and explore technically feasible solutions to achieve it.

- To value alternative designs and the embedded real options consistently, designers should assume that their willingness to take extra risks should be compensated by a constant amount of expected benefit. These principles can be developed into a transparent simulation-based model without using risk-neutral probabilities. The model is easily extendable to multiple sources of uncertainty.
- Consistent valuation reveals two important effects of embedding flexibility and real options in a design. One is obviously the potential for higher values. A more subtle but very important effect is a spectacular change in the effect of risk on the venture.

REFERENCES

Archambeault, W. 2002. For lease or sale: Telecom hotels. *Boston Business Journal.* http://boston.bizjournals.com/boston/stories/2002/09/02/story2.htm.

Baldwin, C., and Clark, K. 2000. *Design rules I: The power of modularity.* Cambridge, MA: MIT Press.

Bengtsson, J. 2001. Manufacturing flexibility and real options: A review. *International Journal of Production Economics* 7:213–224.

Biegler, L. T., Grossman, I. E., and Westenberg, A. W. 1997. *Systematic methods of chemical process design.* Upper Saddle River, NJ: Prentice Hall.

Brennan, M., and Schwartz, E. 1985. Evaluating natural resource investments. *Journal of Business* 58(2): 135–157.

Carr, P. 1988. The valuation of sequential exchange opportunities. *Journal of Finance* 43(5): 1235–1256.

Cullimore, B. 2001. Beyond point design evaluation. *ASME Fluids Engineering Division,* summer meeting. http://www.crtech.com/docs/papers/NewOsummary.pdf.

Décamps, J., Mariotti, T., and Villeneuve, S. 2004. Irreversible investment in alternative projects. In *8th Annual Real Options Conference,* Montreal, Canada.

de Neufville, R., Scholtes, S., and Wang, T. 2006. Valuing options by spreadsheet: Parking garage case example. *Journal of Infrastructure Systems* 12(2): 107–111.

de Weck, O., Chaize, M., and de Neufville, R. 2003. *Enhancing the economics of communications satellites via orbital reconfigurations and staged deployment.* Reston, VA: American Institute of Aeronautics and Astronautics.

Geltner, D., Riddiough, T., and Stojanovic, S. 1996. Insights on the effect of land use choice: The perpetual option and the best of two underlying assets. *Journal of Urban Economics* 39:20–50.

Greden, L., and Glicksman, L. 2004. Architectural flexibility: A case study of the option to convert to office space. *8th Annual Real Options Conference.*

Kalligeros, K. 2006. Platforms and real options in large-scale engineering systems. PhD thesis, MIT.

Kalligeros, K., and. de Weck, O. 2004. Flexible design of commercial systems under market uncertainty: Framework and application. *10th AIAA/ISSMO Multidisciplinary Analysis and Optimization Conference,* AIAA-2004-4646.

Kester, W. C. 1993. Turning growth options into real assets. In *Capital budgeting under uncertainty,* ed. R. Aggarwal, pp. 187–207. Englewood Cliffs, NJ: Prentice Hall.

Kogut, B., and Kulatilaka, N. 1994. Options thinking and platform investments: Investing in opportunity. *California Management Review* 36:4.

Kulatilaka, N. 1993. The value of flexibility: The case of the dual-fuel industrial steam boiler. *Financial Management* 13(3): 271–280.

Majd, S., and Pindyck, R. 1987. Time to build, option value, and investment decisions. NBER working papers 1654. Cambridge, MA: National Bureau of Economic Research. http://ideas.repec.org/p/nbr/nberwo/1654.html.

Margrabe, W. 1978. The value of the option to exchange one asset for another. *Journal of Finance* 33(1): 177–186.

Markish, J., and Willcox, K. 2003. A value-based approach for commercial aircraft conceptual design. *AIAA Journal* 41(10): 2004–2012.

McDonald, R., and Siegel, D. 1986. The value of waiting to invest. *Quarterly Journal of Economics* 101:177–186.

Myers, S. C., and Majd, S. 1990. Abandonment value and project life. *Advances in Futures and Options Research* 4:1–21.

Rayman, S., and Zwecher, M. 1997. Monte Carlo estimation of American call options on the maximum of several stocks. *Journal of Derivatives* 5(1): 7–23.

Reinertsen, D. 1997. *Managing the design factory*. New York: Free Press.

Sethi, A. K., and Sethi, S. P. 1990. Flexibility in manufacturing: A survey. *The International Journal of Flexible Manufacturing Systems* 2(4):289–328.

Suh, E. 2005. Flexible product platforms. PhD thesis, MIT, Department of Aeronautics and Astronautics, Cambridge, MA.

Tannous, G. 1996. Capital budgeting for volume flexible equipment. *Decision Sciences* 27(2): 157–184.

Triantis, A., and Hodder, J. 1990. Valuing flexibility as a complex option. *Journal of Finance* 49(2): 545–565.

Vollert, A. 2003. *A stochastic control framework for real options in strategic valuation*. Boston: Birkhauser.

Zhao, T., and Tseng, C. 2003. Valuing flexibility in infrastructure expansion. *Journal of infrastructure systems* 9(3): 89–97.

11 Real Options Model for Workforce Cross-Training

David A. Nembhard
Pennsylvania State University

Harriet Black Nembhard
Pennsylvania State University

Ruwen Qin
Missouri University of Science and Technology

CONTENTS

Workforce cross-training involves a dynamic investment on workforce flexibility. In this chapter, we propose a real options framework that models the cross-training policy as an approximation of an American call option using binomial lattices. Value stems from the merit of dynamic cross-training compared with the deterministic case using traditional discounted cash flow techniques. This work is discussed in the context of a volatile production system characterized by product dynamics, labor dynamics, task heterogeneity, and workforce heterogeneity. Results suggest that cross-training based on the real options approach is dependent on the production capability and the level of workforce heterogeneity. Thus, valuing workforce flexibility using real options has strategic utility beyond that of the net present value approach.

11.1 CROSS-TRAINING TO ACHIEVE WORKFORCE FLEXIBILITY

In order to survive in today's competitive and volatile economic environment, many organizations are trying to increase their opportunities by adopting competitive strategies such as flexible manufacturing. Cross-training, one of the methods used to increase the capability of production systems by enhancing workforce flexibility, is becoming more widely used to meet such challenges as competition intensifies. Cross-training can benefit production systems by enhancing the production profit and compensating the potential loss of production due to volatilities of the workforce and products. However, the decision to cross-train should be considered carefully because there are many complex and uncertain factors including labor dynamics, product dynamics, task heterogeneity, and worker heterogeneity. Clearly, training workers regardless of these uncertain factors in production systems may cause deviations from investors' expectations by putting the investor at risk of losing irreversible investment capital (Pindyck 1988). An efficient approach for making decisions on workforce flexibility investment is indispensable for managers in a competitive environment. The motivation for our work stems from meeting this need.

In traditional discounted cash flow (DCF) techniques, the net present value (NPV) of an investment project is calculated for a stream of expected net cash flows at a "risk-adjusted" discount rate that reflects the risk of those cash flows. Fama and French (1997) emphasized the difficulties of estimating risk-adjusted discount rates. Nevertheless, for this method, the investment decision is made without regard for uncertainty in the production environment and without regard for the idea that projects at different phases represent a sequence of independent investment opportunities. An immediate accept or reject decision is then made by accepting all projects with a positive NPV. This is the deterministic managerial strategy because at the outset the manager adopts an irrevocable investment policy from which she or he cannot depart regardless of whether the future remains faithful to or deviates from the initially expected scenario. In the absence of a dynamic investment policy, the probability distribution of the NPV would be reasonably symmetric. However, a flexible workforce can adapt its future actions depending on the future production environment. This aspect introduces an asymmetry in the probability distribution of the NPV that expands the value of investment opportunity by improving its upside potential, while limiting downside losses relative to managers' initial expectations under the deterministic investment policy. The managerial strategic value of various projects cannot be properly captured by traditional DCF techniques due to their dependence on future events that are uncertain at the time of the initial decision.

In direct contrast, a real options framework has the potential to capture the essence of such a situation properly. It is apparent that increasing the flexibility of production systems by cross-training is one operational option. That is, managers can decide to cross-train when needed. Cross-training represents a dynamic investment policy on workforce flexibility that parallels the concept of real options theory and correspondingly may be associated with option values. The goal of this chapter is to develop a model to financially value cross-training and quantitatively evaluate the proposed model in a volatile production system that has various sources of uncertainty and where on-the-job cross-training is available.

11.2 CROSS-TRAINING IN A SEQUENTIAL PRODUCTION SYSTEM

A number of researchers have quantified the costs and benefits of cross-training, including Gerwin (1993), Ebeling and Lee (1994), and Molleman and Slomp (1999). Van den Beukel and Molleman (1998) identify the volatility and the uncertainty in the market as the main reasons for an organization to invest in cross-training programs. Nembhard and Norman (2007) measure the effect of cross-training in terms of worker efficiency and responsiveness in the context of several important workplace factors including product, manufacturing process, and workforce dynamics, each of which can cause significant disruptions in productivity. The cost aspects of cross-training offer a basis for work in this chapter wherein cross-training may lower current production efficiency temporarily as workers become less specialized during cross-training, but have later advantages in meeting demand for future products or services. That is, cross-training is a risky investment in workforce flexibility with an opportunity cost. A dynamic investment policy is appropriate because it is undesirable to pay for the extra capability of workforce flexibility if it is not needed, or if the profit from cross-training is too limited to cover the cost of the investment. This point of view leads us to model cross-training using a real options framework.

In order to illustrate this problem, we consider a simple sequential production system as shown in Figure 11.1 in which each station involves a unique separate task. We assume that work in process (WIP) is allowed to accumulate without limit between each of the adjacent stations and is available for immediate processing by the next station. Only after a unit has moved through all workstations is it considered finished.

Each workstation requires exactly one worker, so two workers are involved in each sequential production line. We assume that prior to cross-training, one worker specializes on the base task, and the other specializes on the complex task. However, if a worker is absent from the line, the other one will cover both tasks. During cross-training, each worker not only works on the assigned task but also learns the other, less familiar task. They change positions according to a station rotation interval that we set as a constant when cross-training. After cross-training, workers are skilled at all the tasks, and they do not lose their skills.

We can represent the output of individual workers using a hyperbolic learning model from Molleman and Slomp (1999):

$$y(t) = m_k \frac{t + m_p}{t + m_p + m_r} \tag{11.1}$$

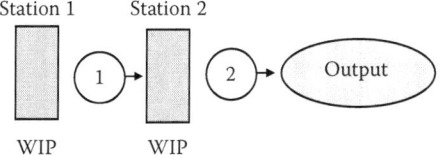

FIGURE 11.1 Sequential production line.

where $y(t)$ is the productivity rate corresponding to t time periods of work ($t \geq 9$), m_k is the fitted parameter estimating the steady-state productivity rate, m_p is the fitted parameter representing initial expertise, m_r is the number of time periods of work and initial expertise required to get halfway to m_k, and $m_p + m_r > 0$.

There are four factors that may affect the model in Equation (11.1) that are described as follows. The first factor is task heterogeneity. We set the complexity level of the *base task* to be 1. We standardize the task complexity level of the *complex task* correspondingly. To model the task heterogeneity in the production line, we let the task complexity level of the complex task be twice that of the base task.

The second factor is worker heterogeneity, which represents the fact that capabilities can differ substantially among workers (Campbell 1999). To model worker heterogeneity, we assume that an *average worker* has an average level of capability, based on the set of workers in the pool. A second *random worker* is sampled from the worker pool to complete the team for a sequential production line. We define $m_{\bullet R \bullet}$ to represent randomly selected worker parameters, $m_{\bullet A \bullet}$ as average worker parameters, $m_{\bullet\bullet B}$ as worker parameters on the base task, and $m_{\bullet\bullet C}$ as worker parameters on the complex task. Specifically, this means that m_{kAB} is the steady-state productivity rate of the average worker on the base task (i.e., it is the average value of steady-state productivity rates for the base task of all workers in the work pool), m_{kRB} is the steady-state productivity rate of the random worker on the base task, m_{kAC} is the steady-state productivity rate of the average worker on the complex task, and m_{kRC} is the steady-state productivity rate of the random worker on the complex task. As a means of comparison, we define Δm_k as the deviation of the steady-state productivity rate for each team, and use it as a measure of worker heterogeneity within each team as follows:

$$\Delta m_k = m_{kRB} - m_{kAB} \tag{11.2}$$

The third factor is labor dynamics. Worker absenteeism is likely a stochastic event for a dynamic manufacturing system. Suppose each worker in a team has an equal chance of being absent, and absent for exactly one day per absence. During a worker absence, the remaining worker on that team covers the work of the absent worker for that day. We assume the average absentee rate is once per week, modeled stochastically following a uniform distribution. We assume no worker turnover in this system.

The fourth factor is product dynamics. We assume the product life cycle is a three-month period and the degree of product change is 50 percent. In other words, at the end of each three-month period, one of the two tasks in a sequential production line will be replaced by another one with the same complexity level. On this point, it is reasonable to equate the changed production system to the original situation with specialized workers. Although each worker is not multiskilled, she or he is still skilled at one task. We also assume the training program begins immediately after the product change and is no longer than the product life cycle. When production changes occur, skills acquired on tasks that are removed are lost.

Figure 11.2 illustrates the capability difference of production systems in terms of a productivity rate function such as that in Equation (11.1). The production system with cross-trained workers may have different steady-state productivity than

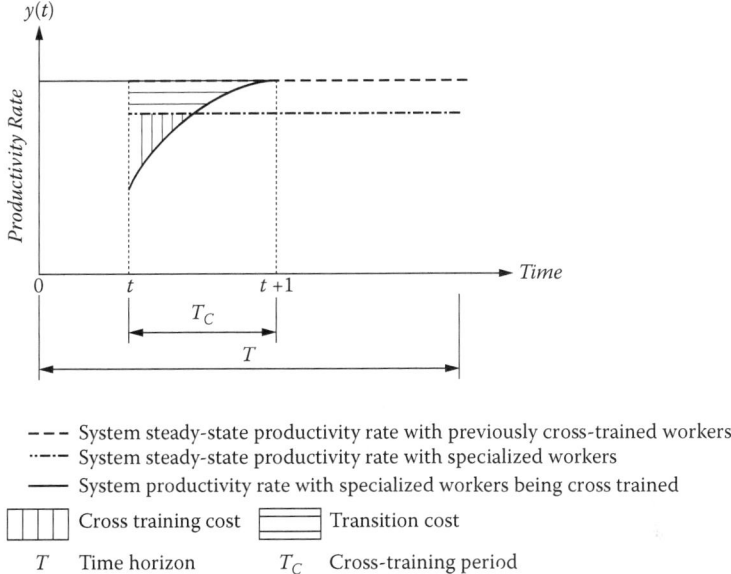

--- System steady-state productivity rate with previously cross-trained workers
··—·· System steady-state productivity rate with specialized workers
——— System productivity rate with specialized workers being cross trained

Cross training cost   Transition cost

T Time horizon T_C Cross-training period

FIGURE 11.2 Production capability under various workforce scenarios.

that with specialized workers. The production system has opportunities to expand its production capability by cross-training. However, this opportunity comes with two costs. One is the cross-training cost, which is the production loss from specialized workers while they are being cross-trained. That is, while they acquire multiple skills through cross-training, they must forgo a portion of production in the short run. The second cost is the system transition cost, which is the unachievable output during the system transitions. In essence, cross-training can potentially expand the future production capability at the expense of some amount of current productivity.

It may be more practical in some cases to treat the overall project as a sequence of subprojects. In Figure 11.3, in each stage, a cost K_t is incurred during cross-training, which is the financial value of production loss while previously specialized workers are cross-trained. However, cross-training in that stage may achieve a portion of the NPV increment of production, Δs_t.

11.3 FORMULATION OF MODEL

We take the view that cross-training can potentially expand future production capability by making a current investment, and is somewhat analogous to an expansion option with associated costs (Kulatilaka and Perotti 1998; Panayi and Trigeorgis 1998). Thus, we parallel this idea to form the expansion option model of cross-training. Since the option to cross-train can be exercised at any time up to the expiration date and since we would like to track the evolution of a dynamic investment in cross-training, we employ an approximation of an American call option.

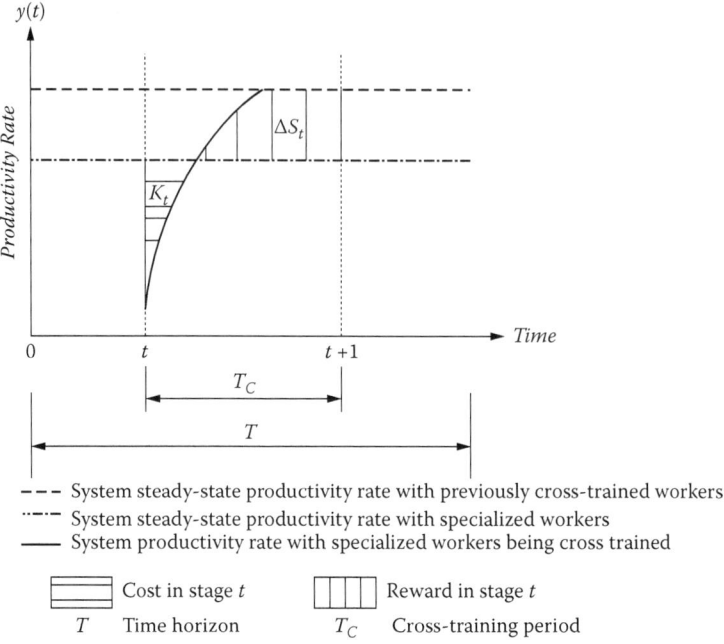

FIGURE 11.3 Systems change in each stage.

We consider the NPV based on traditional DCF techniques for the production system using a specialized workforce and a cross-trained workforce. We then model the extra value obtained by using the real options framework, which allows for a decision to be deferred and/or implemented in stages. In order to use the real options approach to price the workforce flexibility by cross-training, we assume that the net value of the production system during a decision time horizon has a probability distribution. We assume the percentage changes in the output follow an exponential Brownian motion process with a normal distribution, and the uncertain value of the output is perfectly correlated with a portfolio of tradable assets, an approach supported by Black and Scholes (1973).

Let us first establish the NPV (based on traditional DCF techniques) of the production system using a specialized workforce, denoted s_0. This is given by the difference between how much the production system assets are worth (their present value) and how much they cost during a decision horizon T:

$$s_0 = PQ_S - n_W L - CQ_{ST} - F \tag{11.3}$$

where P is the manufacturer's current sales price for unit product; Q_s is the discounted output of a sequential production line during T (at the risk-free rate r) of the stochastic production system with a specialized workforce, n_W is the number of workers in the system, L is the present value of workers' salaries, C is current cost per unit product other than the direct labor cost, and F is the present value of fixed production cost during T.

Now let us establish the NPV of the production system using a cross-trained workforce, denoted NPV_D. Again, this is based on traditional DCF techniques and is given by the maximum of the present value using a cross-trained workforce and the present value using a specialized workforce:

$$NPV_D = \max(Es_0 - K, \ s_0) \tag{11.4}$$

where E is the expansion factor of the production value and K is the present value of cross-training implementation costs. This formulation allows that anytime the value of a cross-trained workforce is less than that of the specialized workforce, then cross-training will not be used. We can further define the expansion factor E as a measure of the NPV by cross-trained workers compared with the NPV by specialized workers:

$$E = \frac{Q_C(P-C)-(n_w L+F)}{Q_S(P-C)-(n_w L+F)} \tag{11.5}$$

where Q_C is the discounted output of a sequential production line during T using a previously cross-trained workforce. That is, cross-training costs have already been incurred, and the workers are at steady state for all cross-trained tasks. We can also define K to be the sum of fixed and variable costs:

$$K = K_p + K_l \tag{11.6}$$

where K_p is the present value of fixed costs of cross-training, and K_l is the present value of variable costs of cross-training, which include both cross-training costs and transition costs. Thus,

$$K_l = (Q_{CC} - Q_{SC})(P - C) \tag{11.7}$$

where Q_{CC} is the discounted output of a sequential production line during a training period T_C using previously cross-trained workers, and Q_{SC} is the discounted output of a sequential production line during a cross-training period T_C using workers being cross-trained. All the present values above are discounted to current time at the risk-free rate.

Note that in Equation (11.4), the cross-training policy is decided entirely at the outset by the maximum value criterion and the NPV_D is obtained accordingly. However, if we frame the cross-training policy using a real options approach, we ultimately can achieve an extra nonnegative value increment relative to NPV_D. This nonnegative value increment is the real options value, V_{RO}:

$$V_{RO} = NPV_{RO} - NPV_D \tag{11.8}$$

where NPV_{RO} is the net present value of production by the real options approach (some authors refer to this as the expanded NPV, or ENPV). In other words, since NPV_{RO} incorporates flexibility and adaptability and exceeds the NPV_D, the V_{RO} indicates the value of workforce flexibility by cross-training using the real options approach compared with cross-training using traditional DCF techniques.

Binomial lattices, approximations of call options, are used to model the NPV_{RO} (Cox et al. 1979). Two binomial lattices are needed: the first is the lattice of the underlying production value, while the second is the options valuation lattice. We suppose that the option lasts for time T, and that the time interval ΔT is the time scale between steps. (For example, if an option has a one-year maturity span and the binomial lattice has ten steps, then each time-step has a time interval of 0.1 year.) During the life of the option, s_0 can either move up from s_0 to a new level $s_0 u$ or down from s_0 to a new level $s_0 d$, where

$$u = e^{\sigma\sqrt{\Delta T}} \quad \text{and} \quad d = \frac{1}{u} \tag{11.9}$$

and the volatility σ is an annualized value of the NPV of production by specialized workers. We use the annual production volatility σ_Q to determine σ as in Molleman and Slomp (1999). The up (u) and down (d) factors are also related through the risk-neutral rate:

$$p = \frac{e^{r\Delta T} - d}{u - d} \tag{11.10}$$

where r is the risk-free interest rate.

To create the lattice of the underlying production values, we start with s_0, then multiply it by the up (u) and down (d) factors to create up and down branches $s_t^i = s_0 u^i d^{t-i}$ ($i = 0, 1, \ldots, t$), as suggested by Figure 11.4. This bifurcation continues at each node to create a binomial lattice for underlying production values. The intermediate branches all recombine since we are using the risk-neutral rate p. The value of an option is generated by the uncertainties and risk that are captured by the volatility measure σ: the higher the volatility measure, the higher the up factors, and the higher the potential value of an option.

After new information arrives, uncertainty about future cash flows is gradually resolved in the binomial lattice for the underlying variables. Flexibility can then be modeled in the options valuation lattice by allowing varying degrees of flexibility relative to the underlying assets lattice. For the investment problem of cross-training, the option is the ability to expand the financial value of production with associated costs at different stages during the horizon T. We assume the lattice has n stages, and at the beginning of each stage, according to the state, we determine the lattice values, c_i^i. Each is given by the maximum of the value of expansion (i.e., implementing the

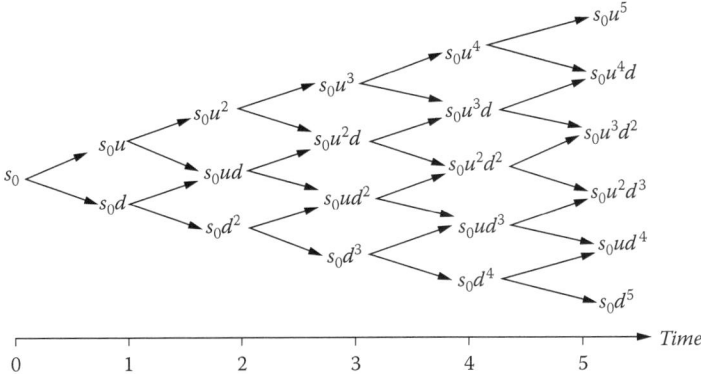

FIGURE 11.4 Lattice of underlying production values.

cross-training policy) and the value of continuation (i.e., keeping the option open to cross-train in the future):

$$c_t^i = \max[Es_t^i - K, c_t^{i*}] \, t = 0, 1, \ldots, n-1 \quad i = 0, 1, \ldots, t \qquad (11.11)$$

We calculate the value of keeping the option open at each intermediate node c_t^{i*} by discounting the one-step-ahead cash flows (c_{t+1}^i and c_{t+1}^{i+1}) backward with risk-neutral rate p:

$$c_t^{i*} = e^{-r\Delta T}[pc_{t+1}^i + (1-p)c_{t+1}^{i+1}], \quad t = 0, 1, \ldots, n-1; \quad i = 0, 1, \ldots, t$$

$$c_t^{i*} = s_t^i, \quad t = n; \quad i = 0, 1, \ldots, t \qquad (11.12)$$

With backward induction in the options valuation lattice, we get the NPV of the option evaluation lattice at time zero, which is the NPV_{RO}.

We may find that the NPV_{RO} is different from the NPV without cross-training, s_0. The difference, Vc, is the financial value of cross-training by the real options approach:

$$V_C = NPV_{RO} - s_0 \qquad (11.13)$$

In other words, Vc compares the value of cross-training versus specialization.

The preceding model can be extended to the state-dependent case wherein the total expansion of the NPV of production is accounted for in the expected NPV increment of production in those stages afterward. Accordingly, the expansion factor previously defined in Equation (11.5) can be redefined as

$$E_t^i = \frac{s_t^i + \Delta s_t^i + E[\Delta s_t^{i*}]}{s_t^i}, \quad t = 0, 1, \ldots, n; \quad i = 0, 1, \ldots, t \qquad (11.14)$$

where $E[\Delta s_t^{i+}]$ is the expected value increment after time t onward due to cross-training, and Δs_t^i is the NPV increment at time t. So, if cross-training at time t, the NPV of production becomes

$$E_t^i s_t^i - K_t^i = s_t^i + E[\Delta s_t^{i+}] + (\Delta s_t^i - K_t^i), \quad i = 0, 1, \dots, t \qquad (11.15)$$

where K_t^i is the state-dependent cost of cross-training for state i and time t. Both Δs_t^i and K_t^i depend on the present value of production at time t and the heterogeneity of the worker team, so under the dynamic investment policy, the managers' decision on cross-training depends on the state for that stage. The expected value increment after time t due to cross-training is written as

$$E[\Delta s_t^{i+}] = e^{-r\Delta T}[p(c_{t+1}^{i+1} - s_t^i u) + (1 - p)(c_{t+1}^{i+1} - s_t^i d)] \qquad (11.16)$$

The third term on the right-hand side of Equation (11.15) may be given by

$$(\Delta s_t^i - K_t^i) = (Q_{SC} - \alpha Q_S)(P - C) \qquad (11.17)$$

where α is a factor used to truncate the exact portion of discounted product belonging to each training period. At the beginning of each stage of the real options valuation lattice, we choose the decision rule that can maximize the NPV of production at that stage:

$$c_t^i = \max[E_t^i s_t^i - K_t^i, c_t^{i*}], \quad t = 0, 1, \dots, n-1; \quad i = 0, 1, \dots, t$$
$$c_t^i = \max[E_t^i s_t^i - K_t^i, s_t^i], \quad t = n; \quad i = 0, 1, \dots, t \qquad (11.18)$$

We remark that, in general, s_t^i and K_t^i may be correlated. In the current formulation, we assume these to be independent. An examination of the correlated case is of interest for future research.

11.4 ANALYSIS OF DECISION TO CROSS-TRAIN IN SIMULATED SEQUENTIAL PRODUCTION SYSTEM

In this section, we illustrate the use of the model and demonstrate its potential use for valuing cross-training. Specifically, we present the results and analysis of the expected financial value of cross-training $E[V_C]$, value of cross-training V_C, and real options value V_{RO}. This leads to an understanding of how cross-training may be implemented to achieve value in a production system. We use a pool of 75 workers, where the attributes of each worker come from empirical data. We model 20 sequential production lines where each line contains two workstations. Each workstation requires one worker, so two workers are involved in each sequential production line. Table 11.1 shows the data of the production system employed in the simulation, wherein we randomly select 20 worker teams where each team constitutes a production line.

TABLE 11.1

Data for Production System

Manufacturer's current sales price for each unit of product (P)	$13.68
Current cost per unit of product other than the direct labor cost (C)	$5.00
Time interval (T)	0.33 year
Time horizon of decision (T)	2 years
Number of workers per unit (n_w)	2
Present value of each worker's salary during the time horizon of decision (L)	$60,000
Present value of fixed production cost during the time horizon of decision (F)	$80,000
Present value of fixed cost of cross-training program rather than output change (K_p)	$40,000
Risk-free rate (r)	0.08
Working hours per year (H)	2080 h (8 h × 5 d × 52 w)

We assume that of the two workers, one works on the base task and the other works on a complex task prior to cross-training. However, if a worker is absent from the line, the other will temporarily cover both tasks. During cross-training, each worker not only works on the assigned task but also learns the other less familiar task. They change positions according to a station rotation interval that we set to one hour when cross-training. After significant cross-training, workers become more skilled at the tasks following the model in Equation (11.1), and do not lose their skills.

Figure 11.5 shows the expected financial value of cross-training, $E[V_C]$, among heterogeneous worker teams. The heterogeneity among these worker teams is

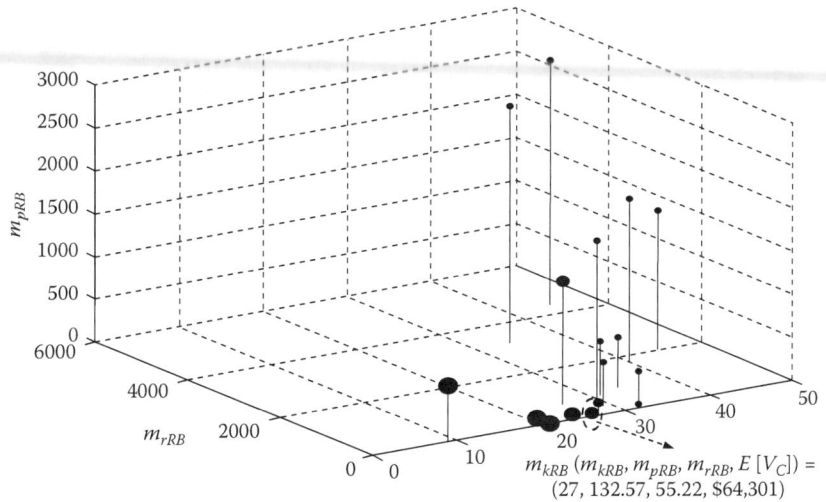

FIGURE 11.5 Expected financial values of cross-training among heterogeneous worker teams.

characterized by the distribution of parameters m_{kRB}, m_{pRB}, and m_{rRB}. The size of each bubble is proportional to the expected financial value of cross-training of that work team. We observe that the variability of the productivity of the worker contributes to the variability of the financial value of cross-training.

To illustrate the calculations in relating the cross-training to the real options valuation, consider the worker team given by the highlighted bubbles in Figure 11.6. For this team, the steady-state productivity values are $m_{kAB} = 29$, $m_{kAC} = 15$, $m_{kRB} = 27$, and $m_{kRC} = 12$ (see Section 11.2); and the discounted output values are $Q_S = 44,112$, $Q_C = 56,430$, $Q_{SC} = 5,810$, and $Q_{CC} = 7,149$ (see Equations (11.5) and (11.7)). With these values, the deterministic NPV of the production system is $NPV_D = \$238,190$ (according to Equation (11.4) with $s_0 = \$182,892$). When $u = 1.248$ and $d = 0.801$, by backward induction in Figure 11.6, the NPV of the production system by the real options approach is $NPV_{RO} = \$246,034$. The real options value is $V_{RO} = \$7,844$ (by Equation (11.8)), and the financial value of cross-training by the real options approach is $V_C = \$63,142$ (by Equation (11.13)). With thirty replications for this team, we ultimately obtain an estimate of the expected financial value of cross-training $E[V_C] = \$64,301$. We perform the analogous replicated simulation for the other nineteen teams.

In Figure 11.7, we use box plots to illustrate the financial value of cross-training V_C, by the real options approach, for the 20 worker teams, sorted by increasing worker heterogeneity, Δm_k. Each box plot stems from the simulation results of thirty replicated experiments. Based on Figure 11.7, there is an advantage to cross-train when the worker with lesser capability is assigned to the complex task. Intuitively, this seems reasonable because cross-training increases flexibility so that each worker in a team has the opportunity to obtain the requisite skills. Given cross-training, the capability of the production system is improved. However, for those worker teams whose higher-capability worker has already been assigned to the complex task, its

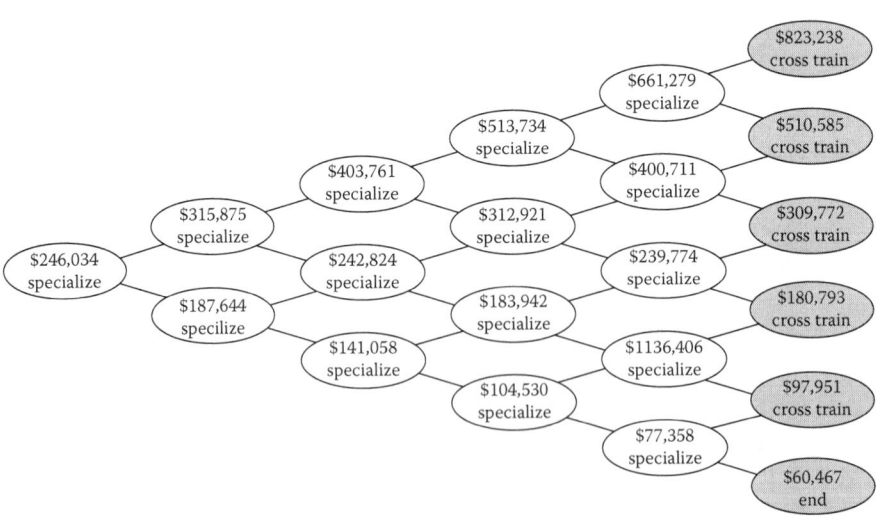

FIGURE 11.6 Real options valuation lattice.

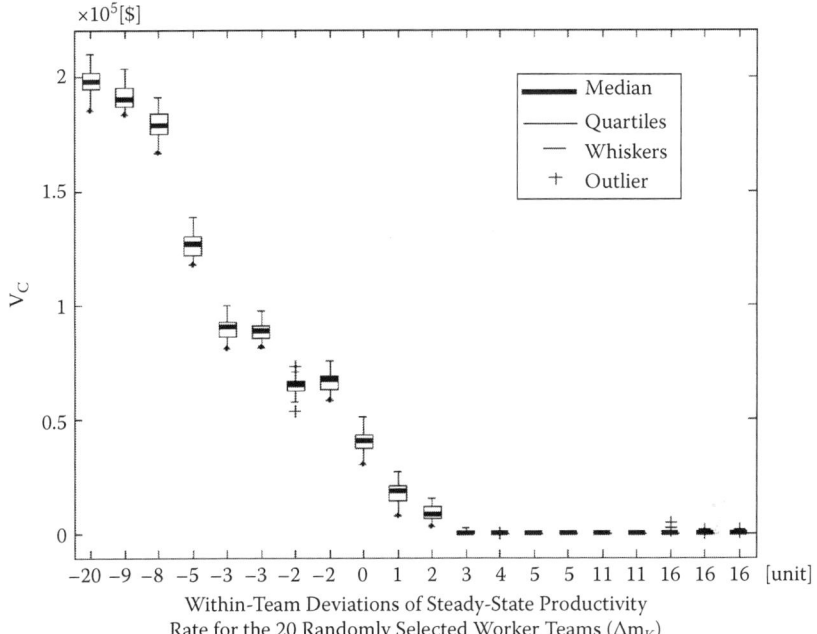

FIGURE 11.7 Financial values of cross-training.

worker assignment is likely to be near optimal. Thus, the expected value of cross-training is near zero for those teams where the randomly selected assignments are already near optimal.

The real options approach provides a nonnegative gain from cross-training and shows how flexibility helps the production system become robust to future uncertainties. From the analysis above, we find the real options framework benefits both the production scenarios and the dynamic investment policy on workforce flexibility.

We also find that less heterogeneous worker teams (lower Δm_k) are associated with higher financial values of cross-training. That is, for the same training level, the greater investment reward should come from those workers with relative lower production capability while being assigned to a more difficult task. This provides insight that may guide further work on optimizing worker assignments and cross-training policies.

Figure 11.8 illustrates the real options values, V_{RO}, for the 20 worker teams, sorted by increasing worker heterogeneity, Δm_k. Each box plot stems from the simulation results of 30 replicated experiments. Since the real options value is the increment of NPV due to a dynamic investment policy, it explicitly shows the difference between the real options valuation of cross-training and that using DCF techniques. In Figure 11.8, real options values mainly exist in those worker teams with the lower-capability workers assigned to the more complex task. For the other teams, real options values are zero in many of the replications. Thus, the real options value, like

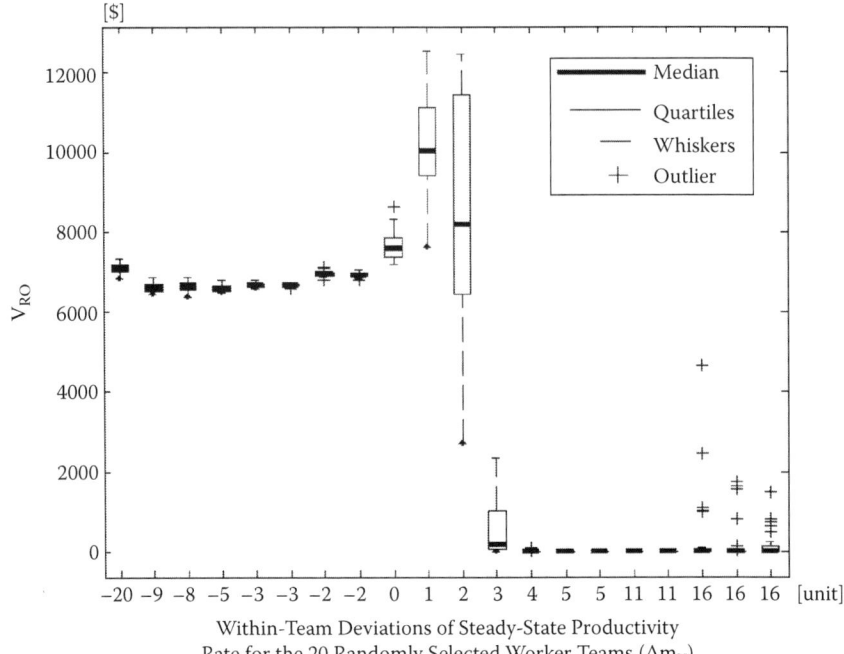

FIGURE 11.8 Real options values.

the value of real options cross-training, is a useful financial measure to identify the optimal level of worker assignment.

Cross-training decisions based on the real options approach can make use of information obtained throughout the production process. Worker teams are often formed arbitrarily at the outset of production, so the deterministic managerial strategy at that time may not be the best given an unknown production scenario. If a dynamic labor factor is involved (e.g., turnover), it may be impossible to get a suitable training policy using a deterministic managerial strategy. If the system is not well designed, a real options approach for cross-training may be beneficial. However, if a system is already near optimal, the real options approach will likely do no better than the DCF technique. So the real options value is effectively pricing the value of information based on the dynamic investment policy for workforce flexibility.

We also find that for those teams whose workers have similar steady-state productivity rates, their real options values have considerable variability. While for those teams whose workers show large differences between steady-state productivity rates, their real options value seems stable and even independent of Δm_k. These indicate that workforce heterogeneity can affect the real options values.

We may extend our insight by reconsidering the simulated production system in Section 11.4 using the state-dependent expansion factor in Equation (11.14). In Figure 11.9, for the real options model using the state-dependent expansion factor and cost, the solution strategy indicates the potential for cross-training at each stage.

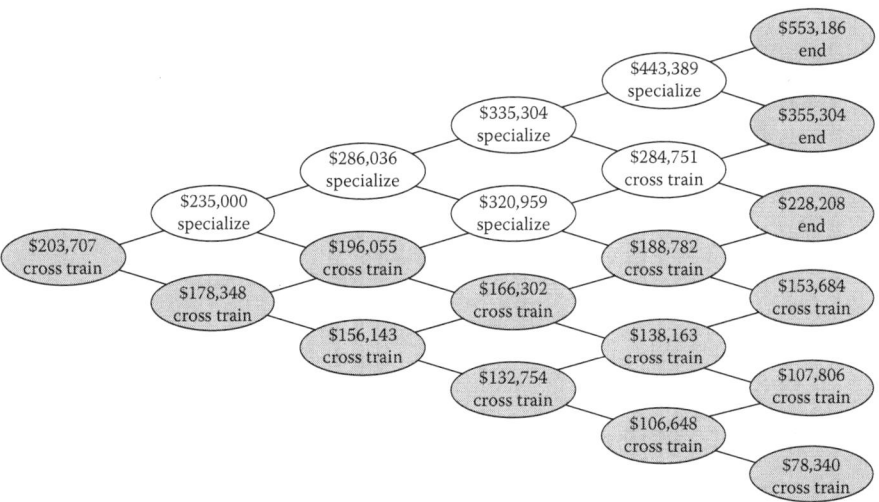

FIGURE 11.9 Real options lattice with state-dependent E_t and K_r.

Cross-training can effectively limit downside losses relative to the manager's initial expectations under the deterministic investment policy.

In Figure 11.10, we illustrate the financial value of cross-training by the real options approach in the state-dependent case, sorted by increasing worker heterogeneity Δm_k. Figure 11.10 is analogous to Figure 11.7, where both models demonstrate the financial value of cross-training in teams with the lower-capability workers assigned to cross-train on the complex task. In comparing these two figures, it is evident that the model with constant expansion factor and cost may be a reasonable approximation of the noncon-stant case, especially for the purpose of pricing the financial value of cross-training, and the state-independent model is somewhat simpler to implement. However, the strategies themselves may be significantly different, as illustrated by Figure 11.6 and Figure 11.9.

In Figure 11.11, we illustrate the real options values, V_{RO}, in the state-dependent case for the 20 worker teams, sorted by increasing worker heterogeneity, Δm_k. Since the invest-ment policy on workforce flexibility is different for these two models, it is not surprising that Figure 11.11 is quite different from Figure 11.8. In Figure 11.11, we find the V_{RO}'s are zero for teams with lower-capability workers assigned to more complex tasks. For these cases, the dynamic investment policy is to always cross-train workers, which is the same as in the deterministic case. Outside of these teams, we can expect a positive V_{RO} in those teams whose workers have similar steady-state productivity rates.

11.5   CONCLUSIONS AND FURTHER RESEARCH

Cross-training can effectively enhance the capability of production systems by increasing the flexibility of the workforce, making them better able to respond to changes in demand and the competitive environment. However, there may be signifi-cant costs involved with acquiring this flexibility.

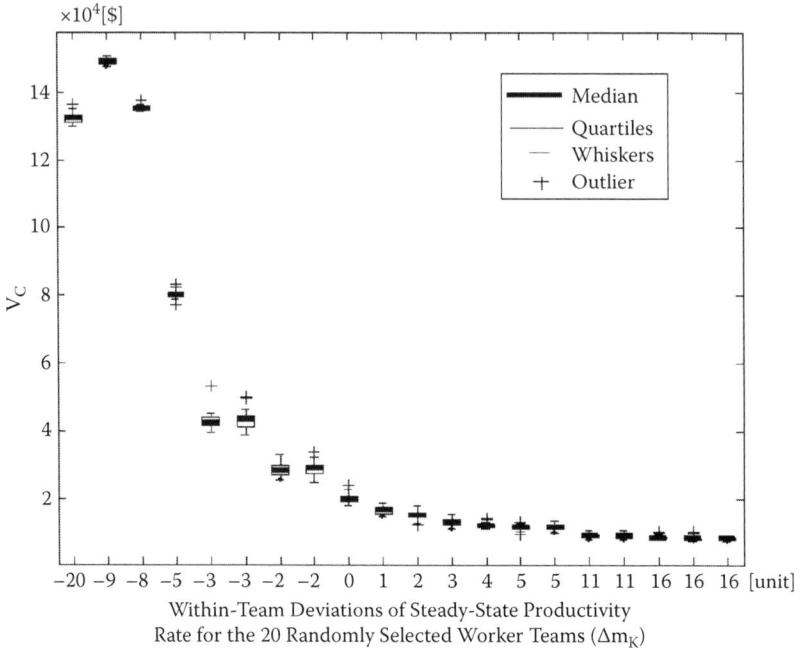

FIGURE 11.10 Financial values of cross-training with state-dependent E_t and K_t.

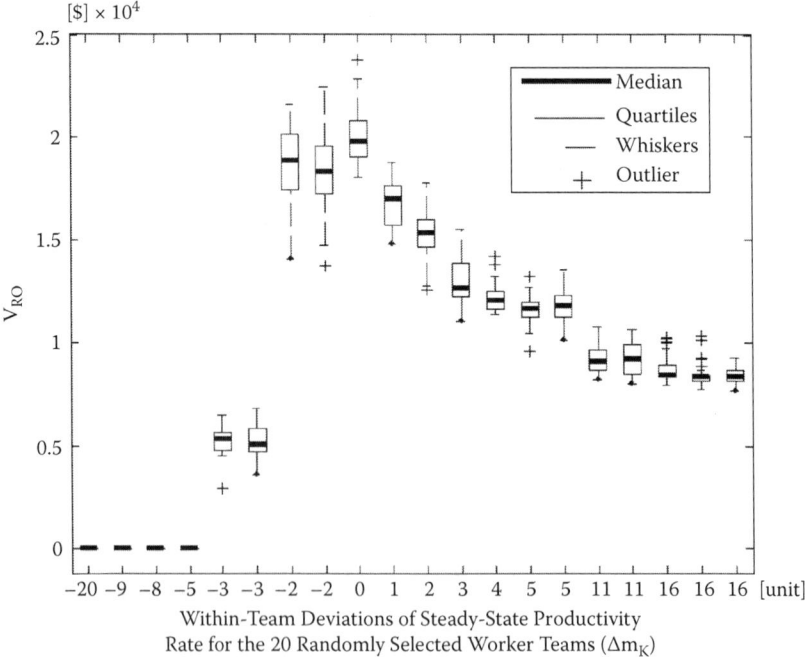

FIGURE 11.11 Real options values with state-dependent E_t and K_t.

In this chapter, we presented a model for cross-training using a real options framework. We analyzed a cross-training policy in a stochastic production system with typical workplace factors including production processes, product dynamics, workforce dynamics, and task and worker heterogeneities. We also modeled and examined both constant and state-dependent expansion factors. With this approach, the cross-training policy can adapt to current production scenarios at different phases, and correspondingly managers can decide to train workers only when there is a likelihood that some profit from cross-training can be achieved. Cross-training can potentially meet the requirements for system improvement while avoiding an unwise investment scenario.

The dynamic cross-training policy provides a useful schedule that satisfies the investment requirement of workforce flexibility. This illustrates the feasibility of the real options framework not only for valuing cross-training but also for operational investment. An important extension of this work is to focus on applications of the flexible managerial strategy in production system control. Future work may include identifying additional information in production processes that can be used by these models.

ACKNOWLEDGMENTS

The authors would like to acknowledge the support of the National Science Foundation, for the first author under SES-0217666, and for the second author under grant DMI-0217924. We also recognize the contributions of Ayse Gurses on an earlier working paper related to this research. Finally, we would like to express our thanks to the referees for their helpful suggestions and comments.

REFERENCES

Black, F., and Scholes, M. 1973. The pricing of options and corporate liabilities. *Journal of Political Economy* 81(3): 637–654.

Campbell, G. M. 1999. Cross-utilization of workers whose capabilities differ. *Management Science* 45(5): 722–732.

Cox, J. C., Ross, S. A., and Rubinstein, M. 1979. Option pricing: A simplified approach. *Journal of Financial Economics* 7(3): 229–263.

Ebeling, A. C., and Lee, C. Y. 1994. Cross-training effectiveness and profitability. *International Journal of Production Research* 32(12): 2843–2859.

Fama, E. F., and French, K. R. 1997. Industry costs of equity. *Journal of Financial Economics* 43(2): 153–193.

Gerwin, D. 1993. Manufacturing flexibility: A strategic perspective. *Management Science* 39(4): 395–410.

Kulatilaka, N., and Perotti, E. C. 1998. Strategic growth options. *Management Science* 44(8): 1021–1031.

Molleman, E., and Slomp, J. 1999. Functional flexibility and team performance. *International Journal of Production Research* 37(8): 1837–1858.

Nembhard, D. A., and Norman, B. A. 2007. Cross training in production systems with human learning and forgetting. In *Workforce cross training,* ed. D. Nembhard, 111–129. Boca Raton, FL: CRC Press.

Panayi, S., and Trigeorgis, L. 1998. Multi-stage real options: The cases of information technology infrastructure and international bank expansion. *The Quarterly Review of Economics and Finance* 38:675–692.

Pindyck, T. S. 1988. Irreversible investment, capacity choice, and the value of the firm. *American Economic Review* 78(5): 969–985.

Van den Beukel, A. L., and Molleman, E. 1998. Multifunctionality: Driving and constraining forces. *Human Factors and Ergonomics in Manufacturing* 8(4): 303–321.

12 Real Options Design for Sustainable Product Quality Management

Julie Ann Stuart Williams
University of West Florida

Mehmet Aktan
Atatürk University

Harriet Black Nembhard
Pennsylvania State University

CONTENTS

Sustainable products are receiving significant attention worldwide from customers, industries, and governments. Companies that can efficiently respond with more sustainable products may recover more long-term value from their operations. Flexibility measures response value in an environment of price uncertainty. We consider a company that has the flexibility of producing both ordinary and more sustainable "green" products, and is striving to improve its overall quality. We present a model to evaluate the optimal strategies that will maximize the expected profit using real options analysis. We illustrate the use and sensitivity of the model for desktop computer production. Our approach gives decision makers a way to choose the appropriate strategies to maximize the expected profit when there is flexibility of producing ordinary and green products under price uncertainty and switching costs between strategies. We discuss future sustainable product quality attributes and their product life cycle management implications for the electronics industry.

12.1 SUSTAINABLE PRODUCTION IN FLEXIBLE SYSTEMS

Recent reports show that corporations are reporting sustainable development activities at an increasing rate (KPMG International 2005). Consequently, research and development efforts have produced new tools to help manufacturers consider environmental attributes in product design (see, e.g., Simon 1992; Stuart 2001; Mangun and Thurston 2002; Tien et al. 2005; Donnelly et al. 2006; Rossi et al. 2006; Ny et al. 2006; Byggeth et al. 2007; Boks and Stevels 2007). Accordingly, so-called green products that are more sustainable are receiving significant attention from customers, industries, and governments around the world. However, there remains a gap between the increasing number of companies reporting awareness of sustainable development principles and the impact on product design and operations (Rondinelli 2007).

Another indicator of the focus on environmental quality is the emergence of the International Standards Organization (ISO) 14000 protocol, which specifies environmental management and standards and promotes sustainable development. While ISO 14000 builds upon the foundation of ISO 9000 for quality management to satisfy customer requirements, it also seeks to satisfy the general public (Morris 2004). An integrated set of requirements for both ISO 9001 and ISO 14001 provides organizations with the opportunity to prepare documentation for the combined system standards (Puri 1996). The ISO reports that its ISO 9000 and ISO 14000 standards are implemented by over 1 million organizations in 161 countries (ISO 2008).

From a management strategy viewpoint, it is important for an organization to determine when it can benefit from responding to the market's changing demand for more sustainable "green" products. In a complex system, one necessity is the ability to react to events as they unfold. Thus for the organization, as the demand for a particular green product increases or decreases, it becomes obvious that the flexibility to modify production must hold some value. In this chapter, we apply the real options approach to determine the optimal strategies between green and ordinary production, and we estimate the expected value of such flexibility over ordinary production. When the production system is flexible, the company can switch production mode among ordinary products, green products, and a combination of both products. The key advantage and value of real options analysis are to integrate managerial flexibility into the valuation process and thereby assist in making the best decisions (Dixit and Pindyck 1994; Trigeorgis 1999; Amram and Kulatilaka 1999; Schwartz and Trigeorgis 2001; Brach 2003). Motivated in part by the attention to joint implementation or the integration of ISO 9000 and ISO 14000 (see, e.g., Prakash 1999; Renzi and Cappelli 2000; Strachan et al. 2003), we incorporate the aspect of ongoing quality control and improvement using a time series model. We demonstrate that by using the real options approach, a company can determine the optimal strategies and the resulting maximum value of flexibility in a system.

12.2 SUSTAINABLE PRODUCT QUALITY MANAGEMENT ISSUES IN DESIGN AND MANUFACTURING IN THE ELECTRONICS INDUSTRY

As noted in the introduction, green products are emerging from the demand-pull of many constituents with new attitudes toward environmental values (Ottman 1992). In this chapter, we specifically consider the supply chain for desktop computers, which represents a significant global industry. Worldwide sales for personal computers, which include desktop and portable computers, exceeded 230 million in 2007 (Gartner 2008; IDC 2008). Product design decisions impact this supply chain, which begins with material choice, then flows to the supplier, manufacturer, retailer, and consumer; the supply chain is extended with a demanufacturing infrastructure to recycle materials (see Figure 12.1). Demanufacturing may include product remanufacturing, component recovery, and/or materials recycling (Maslennikova and Foley 2000; Williams 2006).

Many factors affect the environmental quality of the product across this supply chain. A green product design may include attributes such as recycled material content (U.S. Environmental Protection Agency [EPA] 2005; Qu et al. 2006; Williams 2007) or reduced energy consumption during use (Energy Star 2005; Tuite 2006; IVF 2007; Aoe 2007). The EPA Energy Star Computer Program has been a catalyst for hardware and software design that includes component and circuit configuration, intelligent sensors, and power management software (Eren 2004; Perry 1995; Lapujade and Parker 1995; Liu 1994). An added benefit to designing for lower equipment power consumption is increased reliability from reduced heat dissipation (Bensen 1996).

The determinants of the consumer's purchase decision for recycled products may include the price, believed quality, and psychological benefit (Bei and Simpson 1995). A quality-based model has been developed to analyze the strategic demand and supply as well as policy issues concerning the development of products with conflicting traditional and environmental attributes (Chen 2001). Stuart and Sommerville (1998) summarized the literature for life cycle design guidelines for six specific life cycle characteristics. Since some key life cycle parameters may be uncertain, several

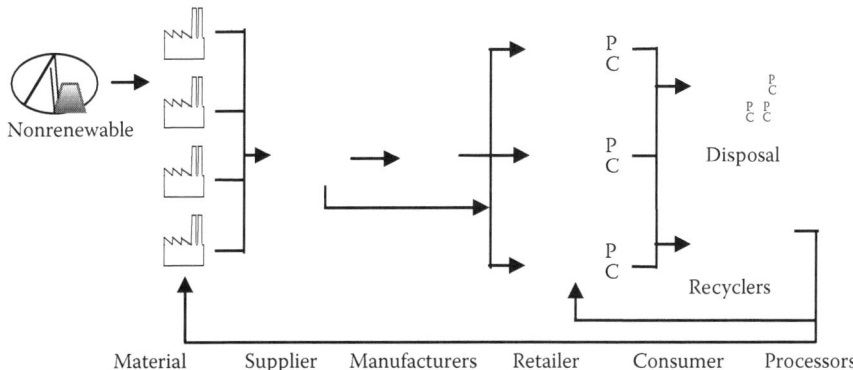

FIGURE 12.1 Current supply chain for desktop computers.

authors propose Monte Carlo simulations to evaluate the sensitivity of parameters in their design tools, which consider both costs and wastes at the material and component acquisition, primary and secondary assembly, and disassembly stages of the product life cycle (Bras and Emblemevag 1996; Sandborn and Murphy 1999). A structured methodology for formulating end-of-life strategies using specific examples from consumer electronics products is presented in Rose et al. (2002). Quotes for environmentally weighted recyclability (QWERTY) is a more comprehensive approach to evaluate a product design by determining an environmentally weighted recycling score in comparison to a weight-based recycling score (Huisman et al. 2003).

At present, product design evaluation tools that consider sustainability factors continue to evolve, but primarily involve checklists and environmental impact scoring (Keoleian and Menerey 1994; Alting and Legarth 1995; Stuart and Sommerville 1998; Lenox et al. 2000; Byggeth and Hochschorner 2006; Ge and Wang 2007). These approaches can effectively compare the outcome of a product in the absence of uncertainty, but offer no means by which to design products for global sustainability in the face of uncertainty. Deterministic design and infrastructure decision tools provide opportunities for sensitivity analysis but do not directly address uncertainty (Clegg et al. 1995; Stuart et al. 1999; Bennett and Yano 2004).

In our work, we integrate real options into the supply chain framework for more sustainable electronic products in order to enhance the understanding of the value of flexibility regarding decision making in this domain context.

12.3   REAL OPTIONS FOR THE FLEXIBLE SUPPLY CHAIN

In the present work, to evaluate the value of flexibility to enhance the sustainability of product manufacturing operations, it is useful to view system control as the ability to switch between states of operation (e.g., corresponding to different combinations of supplier levels, product mixes, and processors). These states may be discrete (e.g., an open or closed plant) or continuous (e.g., operating at different levels of capacity). From this viewpoint, the value of the sustainable materials' quality, energy consumption, or waste generation implications of product design and product life cycle management decisions will typically depend on the underlying uncertainties in a nonlinear fashion. In the real options literature, there are various examples of switching between different operating states (as in Brennan and Schwartz 1985; Triantis and Hodder 1990; Hodder and Triantis 1993).

Two classes of models arise from the option to switch between states over time: when a cost *does not* occur due to a switch in the decision and when a cost *does* occur. Real options without switching costs have the distinguishing feature of time separability, making them easier to value. These real options can be treated as a bundle of European options (i.e., one that can be exercised only on the expiration date) with different maturities and then valued using numerical procedures such as the binomial lattice (Cox et al. 1979; Boyle et al. 1989), the pentanomial lattice (Kamrad and Ritchken 1991; Nembhard et al. 2002), or Monte Carlo simulation (Hull 2008; Nembhard et al. 2003). Valuing a real options problem with switching costs is, of course, more difficult than valuing one without switching costs. Switching costs capture the additional expenditure to change a decision. For example, a switching cost may stem from drawing up a new contract if the manufacturer seeks to change to recycled plastic materials (Xu et al. 2002; Kuswanti et al. 2002; Rios et al. 2003) or from eliminating components and modifying the assembly process if the manufacturer seeks to change the product's power supply design (Directron 2005).

When there are switching costs, options exercised in successive time periods have connections with each other, because a current decision influences the later ones through the extra outlay. In other words, we cannot separate the problem into time periods where the decisions are independent from each other, so we cannot use the feature of time separability. Analytic solutions to real options with switching costs are not attainable if there is more than one state variable; hence, the need for numerical methods arises. Kogut and Kulatilaka (1994) and Huchzermeier and Cohen (1996) presented numerical methods to value multiple options with switching costs. Taudes et al. (1998) proposed to use neural networks to value options with switching costs, and demonstrated their method by valuing flexibility for a costly production switch among several products. Broadie and Glasserman (BG; 1997) developed a simulation algorithm for estimating the prices of American-style assets. Aktan et al. (2003) combine dynamic programming with the BG approach to develop a simulation procedure to value real options problems with switching costs. Nembhard et al. (2005) take this important step further and address valuation when there is an implementation time lag between the time a decision is made and when it can be implemented. They show that without considering the time lag impact, the value of the operational flexibility could be significantly overestimated.

12.4 FORMULATION OF A MODEL TO EVALUATE THE VALUE OF FLEXIBILITY TO PRODUCE ORDINARY, GREEN, OR BOTH PRODUCT DESIGNS

Specifically, we consider a company that has three production strategies to select from at a number of decision points that employ the status quo or introduce a more sustainable green product design. We assume the company has the *flexibility to switch* between the three strategies at each decision point. These three strategies are defined as follows.

Strategy 1: Ordinary production. This is the status quo strategy where only an ordinary product design is employed to make a quantity of N items per time interval.
Strategy 2: Green and ordinary production. Under this strategy, part of the capacity is allocated for the more sustainable green product design. The company produces $N = n_o + n_g$ items per time interval, where n_o is the amount of ordinary items, and n_g is the amount of green items produced.
Strategy 3: Green production. Under this strategy, only the more sustainable green product design is employed to make a quantity of N items per time interval.

The strategies are based on the company's material choices and sources, such as recycled versus new, as well as the product design. For example, under the green strategy, the manufacturer may select a cover composed of recycled high-impact polystyrene and a product design with power phase modes to conserve energy.

A company that is concerned with environmental quality may also be concerned with overall quality. Many technical tools of quality improvement may be deployed, such as process monitoring, experimental design, and process control. A goal of such efforts is to drive down the rate of defective items produced, which, in general, can be represented by time series models (Box et al. 1994; Montgomery et al. 1990). We assume that the rate of defective items produced is expected to decrease according to the following first-order autoregressive time series model:

$$z_{ot} = \varnothing_o z_{ot-1} + a_{ot}$$

and

$$z_{gt} = \varnothing_g z_{gt-1} + a_{gt}$$

where z_{ot} and z_{gt} are rates of defective ordinary and green items produced in time interval t, $\varnothing_o$ and $\varnothing_g$ are constant terms representing the reduction factors for the ordinary and green products' defect rate ($\varnothing_o < 1$, and $\varnothing_g < 1$), and a_{ot} and a_{gt} are random shocks.

In order to model the profit over time, we define the following set of parameters and variables:

p_{ot} = price of the ordinary product in time interval t
p_{gt} = price of the green product in time interval t
c_{ot} = unit production cost of the ordinary product in time interval t
c_{gt} = unit production cost of the green product in time interval t
S_{ij} = cost of switching from strategy i to strategy j
σ_o = volatility of the price of the ordinary product
σ_g = volatility of the price of the green product
ρ = correlation for p_{ot} and p_{gt}
r = risk-free interest rate

Prices of the ordinary and the green products, i.e., p_o and p_g, change over time. The company selects a strategy at a number of decision time points. During time interval t, the company makes a profit P_t, which is defined as follows:

$P_t = (p_{ot} - c_{ot})N - c_{ot}z_{ot}N$, if the preceding and current strategies are Strategy 1;
$P_t = (p_{ot} - c_{ot})N - c_{ot}z_{ot}N - S_{i1}$, if the preceding strategy was Strategy i ($i = 2$ or 3), and the current strategy is Strategy 1;
$P_t = (p_{ot} - c_{ot})n_o + (p_{gt} - c_{gt})n_g - c_{ot}z_{ot}n_o - c_{gt}z_{gt}n_g$, if the preceding and current strategies are Strategy 2;
$P_t = (p_{ot} - c_{ot})n_o + (p_{gt} - c_{gt})n_g - c_{ot}z_{ot}n_o - c_{gt}z_{gt}n_g - S_{i2}$, if the preceding strategy was Strategy i ($i = 1$ or 3), and the current strategy is Strategy 2;
$P_t = (p_{gt} - c_{gt})N - c_{gt}z_{gt}N$, if the preceding and current strategies are Strategy 3; and
$P_t = (p_{gt} - c_{gt})N - c_{gt}z_{gt}N - S_{i3}$, if the preceding strategy was Strategy i ($i = 1$ or 2), and the current strategy is Strategy 3.

The goal of the company is to maximize the expected total discounted profit by selecting the appropriate strategies at each decision point, considering the switching costs between strategies. The prices p_o and p_g change over time as

$$dp_o = \mu_o p_o dt + \sigma_o p_o dz$$

and

$$dp_g = \mu_g p_g dt + \sigma_g p_g dz$$

where μ_o and μ_g are drift of the prices for the ordinary and the green products, respectively; σ_o and σ_g are the volatility of the prices of the ordinary and the green products, respectively, and dz is a standard Wiener disturbance term.

Since there are two state variables (i.e., p_o and p_g), we use a multinomial lattice approach to estimate the value of flexibility due to being able to switch between strategies. At each time interval, the two state variables can move up with the rate of u_i, move down with the rate of d_i such that $d_i = 1/u_i$, or stay constant at each time interval. There are five possible movements, as shown in Table 12.1.

Probabilities of movements p_1 through p_5 are given as

$$p_1 = \frac{1}{4}\left\{ \frac{1}{\lambda^2} + \frac{\sqrt{\Delta t}}{\lambda}\left(\frac{\mu_1}{\sigma_1} + \frac{\mu_2}{\sigma_2} \right) + \frac{\rho}{\lambda^2} \right\}$$

$$p_2 = \frac{1}{4}\left\{ \frac{1}{\lambda^2} + \frac{\sqrt{\Delta t}}{\lambda}\left(\frac{\mu_1}{\sigma_1} - \frac{\mu_2}{\sigma_2} \right) - \frac{\rho}{\lambda^2} \right\}$$

$$p_3 = \frac{1}{4}\left\{ \frac{1}{\lambda^2} + \frac{\sqrt{\Delta t}}{\lambda}\left(-\frac{\mu_1}{\sigma_1} - \frac{\mu_2}{\sigma_2} \right) + \frac{\rho}{\lambda^2} \right\}$$

$$p_4 = \frac{1}{4}\left\{ \frac{1}{\lambda^2} + \frac{\sqrt{\Delta t}}{\lambda}\left(-\frac{\mu_1}{\sigma_1} + \frac{\mu_2}{\sigma_2} \right) - \frac{\rho}{\lambda^2} \right\}$$

$$p_5 = 1 - \frac{1}{\lambda^2}$$

TABLE 12.1

Five Possible Movements and Their Probabilities

Change in p_o	Change in p_g	Probability
u_1	u_2	p_1
u_1	d_2	p_2
d_1	d_2	p_3
d_1	u_2	p_4
0	0	p_5

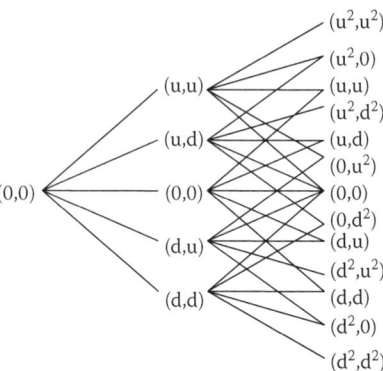

FIGURE 12.2 A pentanomial lattice with two steps.

where Δt is the length of each time interval, ρ is the correlation for the two products' prices, and $\lambda \geq 1$ (Kamrad and Ritchken 1991). For $\lambda = 1$, p_5 is zero, and there are four movements with probabilities p_1 through p_4. The rate of up movement for the ordinary and the green products' prices are given as $u_1 = e^{\lambda \sigma_o \sqrt{\Delta t}}$ and $u_2 = e^{\lambda \sigma_g \sqrt{\Delta t}}$, respectively.

A pentanomial lattice with two steps is shown in Figure 12.2. First and second elements in parentheses represent the changes in the first and second state variables, respectively.

In order to find the optimal set of strategies that maximizes the expected profit, we need to apply a dynamic programming approach in the lattice. If switching between the options is costless, the optimal decision at each node can be analyzed independently. Since we assume that there are switching costs between the options, decisions at the nodes are not independent. When applying the dynamic program, we need to determine the optimal strategy at each node for all immediately preceding possible strategies. The optimal strategies at each node are determined to maximize the expected total profit that will be obtained between that node and the last time interval. Selection of the optimal strategies begins from the last time interval, and moves one time interval at each step until the first node is reached. Then, the set of optimal strategies that maximize the expected total profit for each possible initial strategy can be determined by tracking the set of optimal decisions in the dynamic program.

In each state, the profit is maximized by selecting an option O_t, given that option O_{t-1} was selected in the preceding time interval, and the vector of product prices at time t is e_t. The value V of the total expected profit at time t and state (e_t, O_{t-1}) is defined with the following recursive equation:

$$V_t(e_t, O_{t-1}) = \max_{O_t} E[P_t(e_t, O_{t-1}, O_t)] + e^{-r\Delta t}V_{t+1}(e_{t+1}, O_t)$$

where $E[P_t(e_t, O_{t-1}, O_t)]$ is the expected profit during time interval t.

12.5 REAL OPTIONS ANALYSIS OF ORDINARY AND GREEN DESKTOP PRODUCT DESIGNS

Since two important aspects of green design for computers are material selection and energy consumption during use, we illustrate the use of the real options to evaluate hypothetical ordinary versus green desktop computer designs that differ for these two criteria. For the green product in this example, energy consumption during use is a key design factor that is achieved through hardware and software design (Davis 1994; Liu 1994; Bensen 1996; Gianni et al. 1997; Korn et al. 2004), while material selection includes the use of recycled high-impact polystyrene for the desktop housing (Xu et al. 2002; Kuswanti et al. 2002). Assuming an average plastic desktop computer housing weighs 0.6 kg and that the average price reduction for recycled high-impact polystyrene is $0.30/kg, the production cost may be lowered only 0.002 percent (Rios et al. 2003; Williams et al. 2006). With respect to design modifications to lower energy consumption and heat dissipation (Davis 1994), we assume savings of approximately 6 percent through elimination of the cooling fan with its weight and volume (Directron 2005).

We assume that the levels of the problem parameters at the current time (i.e., $t = 0$) are as follows:

current strategy = Strategy 1
total time horizon = 1 year
number of time intervals = 3
risk-free interest rate (r) = 2 percent
$N = 150,000$ (number of total items to be produced)
$n_o/n_g = 2$ (rate between ordinary and green items produced if strategy 2 is
    selected)
$p_{o0} = \$100$ (price of ordinary product at $t = 0$)
$p_{g0} = \$102$ (price of green product at $t = 0$)
$c_{ot} = 0.90^* \, p_{ot}$ (unit production cost of ordinary product)
$c_{gt} = 0.85^* \, p_{gt}$ (unit production cost of green product)
$\sigma_o = 0.3$ (volatility for the price of ordinary product)
$\sigma_g = 0.3$ (volatility for the price of green product)
$\rho = 0.7$ (correlation for the two products' prices)
$z_{o0} = 3$ percent (rate of defective ordinary products at $t = 0$)
$z_{g0} = 4$ percent (rate of defective green products at $t = 0$)
$\varnothing_o = 0.95$ (discount factor for the number of defective ordinary products)
$\varnothing_g = 0.93$ (discount factor for the number of defective green products)

The design and coordination of hardware and software to achieve energy savings and the identification of sources of recycled high-impact polystyrene require a switching cost (Murphy et al. 2001; Korn et al. 2004). While the power management software options on many desktop computers were not initially activated due to networked device management or software compatibility (Nordman et al. 2000; Christensen et al. 2004; Korn et al. 2004), recent research provides new approaches to effective network power management methods (Harris and Cahill 2005; Gunaratne

TABLE 12.2

Switching Costs between Strategies

		To Strategy		
		1	2	3
From Strategy	1	$ 0	$80,000	$100,000
	2	$ 90,000	$ 0	$ 70,000
	3	$110,000	$80,000	$ 0

et al. 2005; Anand et al. 2005). Switching costs among the three strategies are given in Table 12.2.

Using the above values in a quadrinomial lattice, we can obtain the expected profits and optimal set of strategies at all nodes in the lattice. Figure 12.3 shows the optimal set of decisions that maximizes the expected profit considering all possible levels of ordinary and green products' prices throughout the three time intervals. In each node, there are four elements: the first element shows the preceding strategy, the second element in bold characters shows the strategy that should be applied in the current time interval, the third element shows the up (u) and down (d) movements that have been observed totally in the ordinary product's price, and the last element shows the up or down movements that have been observed totally in the green product's price. If there is a strategy change in a node, that node is colored gray. We see that a switch should be made in the first time interval (from strategy 1 to strategy 3), and a switch back from strategy 3 to strategy 1 should be made at one out of the nine possible nodes in the third time interval.

Figure 12.4 shows the decisions and all possible levels of ordinary and green product prices when there is no flexibility to change the strategy (i.e., strategy 1 is applied and cannot be changed at all three time intervals).

When there is flexibility to produce green and ordinary products, the expected total profit is estimated as $5,433,829. This estimated value is the output of the optimal set of strategies given in Figure 12.3. If we do not have the flexibility to produce

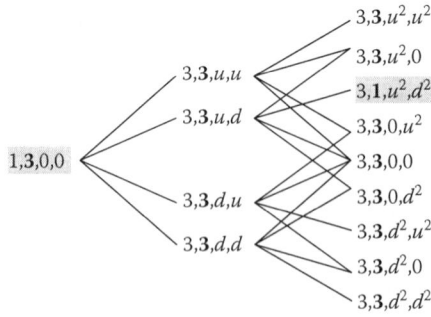

FIGURE 12.3 Optimal switching decisions.

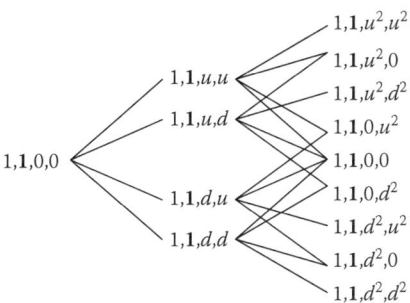

FIGURE 12.4 Single strategy with no flexibility.

green products (i.e., no switching between strategies), the expected profit is estimated as $3,402,407. This estimated value is the output of strategy 1 given in Figure 12.4. Then, the value of green production flexibility is estimated as follows:

$$\text{Estimated value of flexibility} = \$5,433,829 - \$3,402,407 = \$2,031,422$$

In other words, having the green production as an option in addition to the ordinary production provides an expected profit that is $2,031,422 greater than the expected profit of ordinary production. This can be defined as the value of having green production as an additional strategy to the current strategy of ordinary production.

An important part of the valuation analysis is sensitivity of the estimated option value against system parameters. We analyze the behavior of the estimated option value (the value of being flexible to produce the green product) against different levels of the risk-free interest rate (r), the initial price of the ordinary product (p_{o0}), the initial price of the green product (p_{g0}), the demand for the ordinary product (n_o), the demand for the green product (n_g), the price volatility of the ordinary product (σ_o), the price volatility of the green product (σ_g), and the correlation for the ordinary and the green products' prices (ρ).

Figure 12.5 shows the estimated option value against the risk-free interest rate. We see that the estimated option value drops with an increasing rate while the risk-free interest rate increases. This is an expected result since the interest rate has a negative effect on the present value of future cash flows.

Figure 12.6 shows the change of estimated option value against the ordinary and the green products' initial prices. We see that the value of flexibility increases with increasing price of the green product, and it decreases with increasing price of the ordinary product. If the price of the ordinary product is larger than the green product's price, it is more profitable to produce the ordinary product, and, therefore, there is not much value to have a flexibility of producing the green product. On the other hand, flexibility of producing the green product becomes more valuable when the price of the green product increases. As a result, the option value increases with increasing price of the green product.

Figure 12.7 shows the estimated option value against the demand for the ordinary product per time interval. We assume that the total demand for the ordinary and

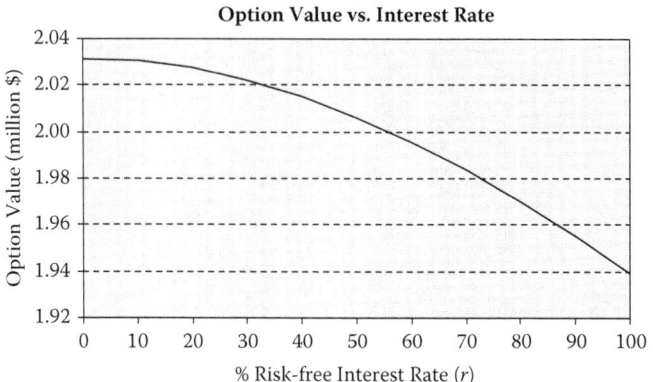

FIGURE 12.5 Estimated option value versus risk-free interest rate.

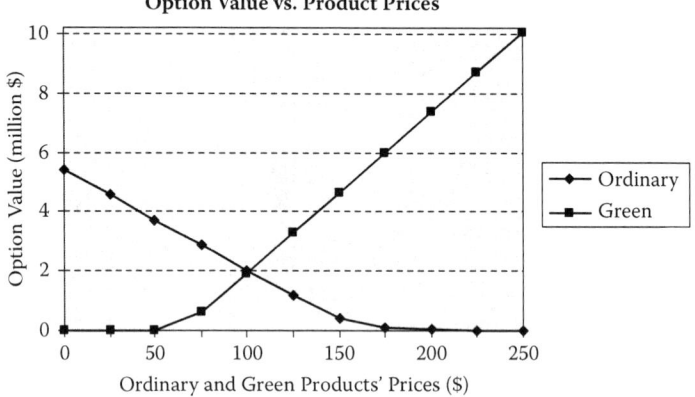

FIGURE 12.6 Estimated option value versus initial prices of the products.

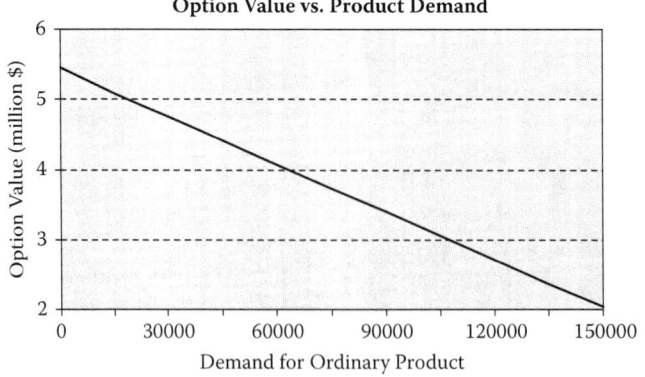

FIGURE 12.7 Estimated option value versus demand for ordinary product.

green products is stable at the level of 150,000 items per time interval. Flexibility of producing the green product has its maximum value when the whole demand is for the green product. The value of flexibility decreases with decreasing demand for the green product.

Figure 12.8 shows the estimated option value against the volatility for the ordinary and the green products' prices. It can be seen that both volatilities have similar small positive effects on the value of flexibility. It can be seen that the value of flexibility is more sensitive to the green product's price volatility when that volatility is between 0.5 and 0.7, and it is more sensitive to the ordinary product's price volatility when that volatility is between 0.7 and 1.0.

Figure 12.9 shows the effect of the correlation for ordinary and green products' prices on the estimated option value. While the correlation increases from −1 to 1, the value of flexibility decreases with a slowing speed. It can be seen that when the price correlation for the ordinary and the green products is negative, there is more value for being flexible to produce the green product.

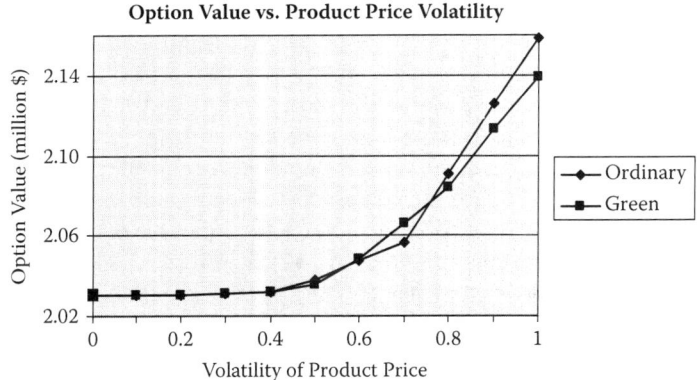

FIGURE 12.8 Estimated option value versus product price volatilities.

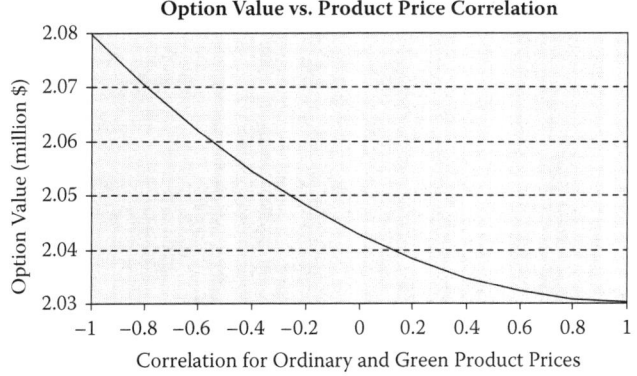

FIGURE 12.9 Estimated option value versus correlation for the product prices.

12.6 SUMMARY

In this chapter, we considered a company that has the flexibility of producing ordinary products and more sustainable green products and is also striving to reduce its overall defective product rate through quality control and improvement efforts. The model we developed for the system used a real options valuation approach to estimate the additional value of flexible production over ordinary production. If the production is flexible, it is assumed that the company is able to switch among three production modes: producing only the ordinary product, producing only the more sustainable green product, or producing both the ordinary and green products.

Using the domain context of desktop computer design, we analyzed the decision to switch among production levels that would allow for better environmental quality. We assumed that switching among the three strategies is costly, and defined the switching costs in a matrix. We estimated the value of green production flexibility that is provided by the three strategies using a real options approach. We presented a complete set of optimal strategies that maximizes the expected total discounted profit and analyzes the sensitivity of the estimated option value against different levels of system parameters.

We believe that our approach shows that decision makers can choose appropriate strategies to maximize the expected profit when there is flexibility of producing ordinary and green products under price uncertainty and switching costs among strategies. For example, this approach could be used to help manufacturers evaluate the use of conventional nonrenewable materials versus biomaterials (Toensmeier 2004, 2007; IVF 2007). Future research opportunities include the interface of this model with product life cycle management and material selection tools. This new capability may help companies close the gap between recognizing the importance of sustainable design and deciding when to pursue the flexibility of producing more sustainable product designs.

ACKNOWLEDGMENTS

This work was supported in part by National Science Foundation grants DMI-0217924 and BES-0124761. Any opinions, findings, and conclusions or recommendations expressed in this chapter are those of the authors and do not necessarily reflect the views of the National Science Foundation.

REFERENCES

Aktan, M., Nembhard, H. B., and Shi, L. 2003. A Monte Carlo simulation approach for valuing real options with switching costs. *EURO/INFORMS Joint International Meeting*, Istanbul, July 6–10.

Alting, L., and Legarth, J. B. 1995. Life cycle engineering and design. *CIRP General Assembly* 44(2): 569–580.

Amram, M., and Kulatilaka, N. 1999. *Real options: Managing strategic investment in an uncertain world*. Boston: Harvard Business School Press.

Anand, M., Nightingale, E. B., and Flinn, J. 2005. Self-tuning wireless network power management. *Wireless Networks* 11:451–469.

Aoe, T. 2007. Eco-efficiency and ecodesign in electrical and electronic products. *Journal of Cleaner Production* 15:1406–1414.

Bei, L. T., and Simpson, E. M. 1995. The determinants of consumers' purchase decision for recycled products: An application of acquisition-transaction utility theory. *Advanced Consumer Research* 22:257–261.

Bennett, D. P., and Yano, C. A. 2004. A decomposition approach for an equipment selection and multiple product routing problem incorporating environmental factors. *European Journal of Operational Research* 156(3): 643–664.

Bensen, P. 1996. Good engineering ensures easy compliance with EPA's Energy Star Program. *Compliance Engineering.* 13(6): 63–68.

Boks, B., and Stevels, A. 2007. Essential perspectives for design for environment: Experiences from the electronics industry. *International Journal of Production Research* 45(18–19): 4021–4039.

Box, G. E. P., Jenkins, G. M., and Reinsel, G. C. 1994. *Time series analysis: Forecasting and control,* 3rd ed. Englewood Cliffs, NJ: Prentice Hall.

Boyle, P. P., Evnine, J., and Gibbs, S. 1989. Numerical evaluation of multivariate contingent claims. *The Review of Financial Studies* 2(2): 241–250.

Brach, M. A. 2003. *Real options in practice.* Hoboken, NJ: John Wiley.

Bras, B. A., and Emblemevag, J. 1996. Activity-based costing and uncertainty in designing for the life-cycle. In *Design for X: Concurrent engineering,* ed. G. Huang. London: Chapman & Hall.

Brennan, M. J., and Schwartz, E. S. 1985. Evaluating natural resource investments. *Journal of Business* 58(2): 135–157.

Broadie, M., and Glasserman, P. 1997. Pricing American-style securities using simulation. *Journal of Economic Dynamics and Control* 21:1323–1352.

Byggeth, S., Broman, G., and Robert, K.-H. 2007. A method for sustainable product development based on a modular system of guiding questions. *Journal of Cleaner Production* 15:1–11.

Byggeth, S., and Hochschorner, E. 2006. Handling trade-offs in ecodesign tools for sustainable product development and procurement. *Journal of Cleaner Production.* 14(15–16): 1420–1430.

Chen, C. 2001. Design for the environment: A quality-based model for green product development: *Management Science* 47(2): 250–263.

Christensen, K., Nordman, B., and Brown, R. 2004. Power management in networked devices. *Computer* 37(8): 91–93.

Clegg, A. J., Williams, D. J. and Uzsoy, R. 1995. Production planning for companies with remanufacturing capability. *Proceedings of the IEEE International Symposium on Electronics and the Environment,* May, Orlando, FL, pp. 186–191.

Cox, J. C., Ross, S. A., and Rubinstein, M. 1979. Option pricing: A simplified approach. *Journal of Financial Economics* 7:229–263.

Davis, J. W. 1994. Fertilizing the green machine. *Information Display* 10(3): 8–11.

Directron. 2005. http://www.directron.com. Accessed February 2005.

Dixit, A. K., and Pindyck, R. S. 1994. *Investment under uncertainty.* Princeton, NJ: Princeton University Press.

Donnelly, K., Beckett-Furnell, Z., Traeger, S., Okrasinski, T., and S. Holman. 2006. Eco-design implemented through a product-based environmental management system. *Journal of Cleaner Production* 14:1357–1367.

Energy Star. 2005. http://www.energystar.gov. Accessed February 2005.

Environmental Protection Agency (EPA). 2005. http://www.epa.gov/epaoswer/non-hw/procure/index.htm. Accessed February 2005.

Eren, H., Al-Shoaili, S., and Milarski, W. 2004. Power awareness and energy efficiency in portable instrument networks. *Proceedings of the ISA/IEEE Sensors for Industry Conference,* August, pp. 96–100.

Gartner, Inc. 2008. http://www.gartner.com/it/page.jsp?id=584210. Accessed February 2008.

Ge, C. P., and Wang, B. 2007. An activity-based modeling approach for assessing the key stakeholders' corporation in the eco-conscious design of electronic products. *Journal of Engineering Design* 18(1): 55–71.

Gianni, R., van der Linden, P., Arreola, O., Chen, H., Goyal, S., Marks, J., Nowshadi, S., and Whatley, T. 1997. Thermal stress relief with power management. *Proceedings of the IEEE International Symposium on Electronics and the Environment,* May, San Francisco, CA, pp. 215–217.

Gunaratne, C., Christensen, K., and Nordman, B. 2005. Managing energy consumption costs in desktop PCs and LAN switches with proxying, split TCP connections, and scaling of link speed. *International Journal of Network Management* 15:297–310.

Harris, C., and Cahill, V. 2005. Power management for stationary machines in a pervasive computing environment. *Proceedings of the 38th Hawaii International Conference on System Sciences,* IEEE Computer Society, Big Island, HI, January 3–6, pp. 285–295.

Hodder, J. E., and Triantis, A. J. 1993. Valuing flexibility: An impulse control framework. *Annals of Operations Research* 45:109–130.

Huchzermeier, A., and Cohen, M. A. 1996. Valuing operational flexibility under exchange rate risk. *Operations Research* 44(1): 100–113.

Huisman, J., Boks, C. B., and Stevels, L. N. 2003. Quotes for environmentally weighted recyclability (QWERTY): Concept of describing product recyclability in terms of environmental value. *International Journal of Production Research* 41(16): 3649–3665.

Hull, J. C. 2008. *Options, futures, and other derivatives,* 7th ed. Upper Saddle River, NJ: Prentice Hall.

IDC, Inc. 2008. http://www.idc.com/getdoc.jsp?containerId=prUS20995107. Accessed February 2008.

International Standards Organization (ISO). 2008. http://www.iso.org/iso/iso_catalogue/management_standards/iso_9000_iso_14000.htm. Accessed February 2008.

IVF Industrial Research and Development Corporation. 2007. Lot 3 personal computers (desktops and laptops) and computer monitors final report (task 1–8). European Commission DG TREN Preparatory Studies for Eco-Design Requirements of EuPs. IVF Report 07004, Molndal, Sweden. http://www.ecocomputer.org. Accessed October 2007.

Kamrad, B., and Ritchken, P. 1991. Multinomial approximating models for options with k state variables. *Management Science* 37(12): 1640–1653.

Keoleian, G. A., and Menerey. D. 1994. Sustainable development by design: Review of life cycle design and related approaches. *Air & Waste* 44:645–668.

Kogut, B., and Kulatilaka, N. 1994. Operating flexibility, global manufacturing, and the option value of a multinational network. *Management Science* 40(1): 123–139.

Korn, D., Huang, R., Beavers, D., Bolioli, T., and Walker, M. 2004. Power management of computers. *Proceedings of the IEEE International Symposium on Electronics and the Environment,* Scottsdale, AZ, May 10–13, pp. 128–131.

KPMG International. 2005. *KMPG international survey of corporate responsibility reporting 2005*. Amsterdam: KPMG Global Sustainability Services.

Kuswanti, C., Xu, G., Qiao, J., Stuart, J. A., Koelling, K., and Lilly, B. 2002. An engineering approach to plastic recycling based on rheological characterization. *Journal of Industrial Ecology* 63–4: 125–135.

Lapujade, P., and Parker, D. S. 1995. Energy savings of an office computer system. *Energy* 20(1): 15–16.

Lenox, M., King. A., and Ehrenfeld, J. 2000. An assessment of design-for-environment practices in leading US electronics firms. *Interfaces* 30(3): 83–94.

Liu, K. H. 1994. Cost-effective desktop computer power management architecture for the Energy Star Computer program. *PESC Record: IEEE Annual Power Electronics Specialists Conference,* 2, 1337–1341.

Mangun, D., and Thurston, D. 2002. Incorporating component reuse, remanufacture, and recycle into product portfolio design. *IEEE Transactions on Engineering Management* 49(4): 479–490

Maslennikova, I., and Foley, D. 2000. Xerox's approach to sustainability. *Interfaces* 30(3): 226–233.

Montgomery, D. C., Johnson, L. A. and Gardiner, J. S. 1990. *Forecasting and Time Series Analysis*. New York: McGraw-Hill.

Morris, A. S. 2004. *ISO 14000 environmental management standards: Engineering and financial aspects*. New York: John Wiley.

Murphy, C. F., Dillon, P. S. and Pitts, G. E. 2001. Economic and logistical modeling for regional processing and recovery of engineering thermoplastics. *Proceedings of the IEEE International Symposium on Electronics and the Environment,* Denver, CO, May, pp. 229–235.

Nembhard, H. B., Shi, L., and Aktan, M. 2002. A real options design for quality control charts. *The Engineering Economist* 47(1): 28–59.

Nembhard, H. B., Shi, L., and Aktan, M. 2003. A real options design for product outsourcing. *The Engineering Economist,* 48(3): 199–217.

Nembhard, H. B., Shi, L., and Aktan, M. 2005. A real options based analysis for supply chain decisions. *IIE Transactions* 37(10): 945–956.

Nordman, B., Meier, A., and Piette, M. A. 2000. PC and monitor night status: Power management enabling and manual turn-off. *Proceedings ACEEE Summer Study on Energy Efficiency in Buildings* 7:7.89–7.99.

Ny, H., MacDonald, J. P., Broman, G. Yamamoto, R., and Robert, K.-H. 2006. Sustainability constraints as system boundaries. *Journal of Industrial Ecology* 10(1–2): 61–77.

Ottman, J. A. 1992. Industry's response to green consumerism. *Journal of Business Strategy.* 13(July–August): 3–7.

Perry, T. S. 1995. The environment. *IEEE Spectrum* 32(1): 60–61.

Prakash, A. 1999. A new-institutionalist perspective on ISO 14000 and responsible care. *Business Strategy and the Environment* 8:322–335.

Puri, S. C. 1996. *Stepping up to ISO 14000: Integrating environmental quality with ISO 9000 and TQM*. New York: Productivity Press.

Qu, X., Williams, J. A. S., and Grant, E. 2006. Viable plastics recycling from end-of-life electronics. *IEEE Transactions on Electronics Packaging Manufacturing* 29(1): 25–31.

Renzi, M. F., and Cappelli, L. 2000. Integration between ISO 9000 and ISO 14000: opportunities and limits. *Total Quality Management* 11(4/5–6): S849–S856.

Rios, P., Stuart, J. A., and Grant, E. 2003. Plastics disassembly versus bulk recycling: Engineering design for end-of-life electronics resource recovery. *Environmental Science and Technology* 37(23): 5463–5470.

Rondinelli, D. A. 2007. Globalization of sustainable development: Principles and practices in transnational corporations. *The Multinational Business Review* 15(1): 1–24.

Rose, C. M., Stevels, A., and Ishii, K. 2002. Method for formulating product end-of-life strategies for electronics industry. *Journal of Electronics Manufacturing* 11(2): 185–196.

Rossi, M., Charon, S., Wing, G., and Ewell, J. 2006. Design for the next generation: Incorporating cradle-to-cradle design into Herman Miller products. *Journal of Industrial Ecology* 10(4): 193–210.

Sandborn, P. A., and Murphy, C. F. 1999. A model for optimizing the assembly and disassembly of electronic systems. *IEEE Transactions on Electronics Packaging Manufacturing* 22(2): 105–117.

Schwartz, E. S., and Trigeorgis, L., eds. 2001. *Real options and investment under uncertainty: Classical readings and recent contributions*. Cambridge, MA: MIT Press.

Simon, F. L. 1992. Marketing green products in the triad. *Columbia Journal of World Business* 27(Fall–Winter): 268–285.

Strachan, P. A., Sinclair, I. M., and Lal, D. 2003. Managing ISO 14001 implementation in the United Kingdom Continental Shelf (UKCS). *Corporate Social Responsibility and Environmental Management* (10): 50–63.

Stuart, J. A. 2001. Clean manufacturing. In *Handbook of industrial engineering*, 3rd ed., ed. G. Salvendy. New York: Wiley-Interscience.

Stuart, J. A., Ammons, J., and Turbini, L. 1999. A product and process selection model with multidisciplinary environmental considerations. *Operations Research* 47(2): 221–234.

Stuart, J. A., and Sommerville, R. S. 1998. A review of life-cycle design challenges. *International Journal of Environmentally Conscious Design & Manufacturing* 7(1): 43–57.

Taudes, A., Natter, M., and Trcka, M. 1998. Real option valuation with neural networks. *International Journal of Intelligent Systems in Accounting, Finance and Management*, 7(1), 43–52.

Tien, S.-W., Chung, Y.-C., and Tsai, C.-H. 2005. An empirical study on the correlation between environmental design implementation and business competitive advantages in Taiwan's industries. *Technovation* 25:783–794.

Toensmeier, P. A. 2004. Markets for biopolymers grow as the materials evolve. *Plastics Engineering* 60(10): 20–21.

Toensmeier, P. A. 2007. Green designs: Many factors determine validity. *Plastics Engineering* 63(10): 54–56.

Triantis, A. J., and Hodder, J. E. 1990. Valuing flexibility as a complex option. *Journal of Finance* 45(2): 549–565.

Trigeorgis, L. 1999. *Real options: Managerial flexibility and strategy in resource allocation:* Cambridge, MA: MIT Press.

Tuite, D. 2006. Multiple standards confound power-supply designers. *Electronic Design* 54(15): 39–44.

Williams, J. A. S. 2006. A review of electronics demanufacturing processes. *Resources Conservation and Recycling* 47:195–208.

Williams, J. A. S. 2007. A review of research towards computer integrated demanufacturing for materials recovery. *International Journal of Computer Integrated Manufacturing* 20(8): 773–780.

Williams, J. A. S., E. Grant, P. Rios, L. Blyler, L. Tieman, L. Duplaga, W. Bonawi-tan, M. Madden, and N. Meyer Guthrie. 2006. Plastic separation planning for end-of-life electronics. *IEEE Transactions on Electronics Packaging Manufacturin* 29(2): 110–118.

Xu, G., Qiao, J., Kuswanti, C., Koelling, K., Stuart, J. A. and Lilly, B. 2002. Characterization of virgin and postconsumer blended high impact polystyrene resins for injection molding. *Journal of Applied Polymer Science* 84:1–8.

13 Real Options in Nanotechnology R&D

Anteneh Ayanso
Brock University

Hemantha Herath
Brock University

CONTENTS

In this chapter, we provide a real options perspective on nanotechnology. We review nanotechnology and its evolutionary progress and, based on real options theory, present a framework for understanding and valuing R&D investment in nanotechnology. We discuss the opportunities and challenges arising from nanotechnology investments from the perspective of real options. More specifically, we attempt to provide a real options-based conceptual framework for firms and governments that strengthens the case for investing in nanotechnology R&D.

13.1 FLEXIBILITY IN RESEARCH AND DEVELOPMENT

Increasingly, organizations are relying on new technologies in order to achieve competitive advantage and sustained growth. Intense market competition and continuous technological innovations are creating enormous business opportunities, but also uncertainties for capital investment decisions regarding research and development (R&D). Evaluating the risks, benefits, and flexibilities associated with an R&D project amid such environments require approaches that are far more comprehensive and effective than traditional methods (Pennings and Lint 1997; Perlitz et al. 1999). The real options approach, which adopts the option pricing framework in financial economics, has received significant attention in recent research into valuing R&D investments and technology risk management (Benninga and Tolkowsky 2002; Herath and Park 1999; Lint and Pennings 1998). The real options approach allows the decision maker to capture and evaluate the wide scope of an application as well as the embedded options in the adoption of an R&D project or a new technology. In other words, the real options approach allows the incorporation of managerial flexibilities into the investment decision as new information becomes available (Huchzermeier and Loch 2001). This capability is particularly important for the effective management of risks as most innovative technologies involve a significant amount of up-front irreversible investments, uncertain costs, and uncertain tangible and intangible benefits that may take several years to realize (Benaroch and Kauffman 1999; Kumar 2002).

13.2   NANOTECHNOLOGY: HISTORY AND EVOLUTION

Nanotechnology refers to the use of science, engineering, and technology to understand and control matter at the atomic or nanoscale level (Guz et al. 2007; Sheetza et al. 2005; Walsh et al. 2008); "nanoscale" refers to dimensions of roughly 1 to 100 nanometers, where a nanometer is one-billionth of a meter. Nanotechnology has been heralded as a promising technology that will contribute to economic prosperity and sustainable development by a broad alliance of policy makers, scientists, and industry representatives (Bhat 2005; Fleischer and Grunwald 2008). In a 1960 seminal talk titled "There Is Plenty of Room at the Bottom," Richard Feynman presented the idea of manipulating and controlling on a small scale:

> A biological system can be exceedingly small. Many of the cells are very tiny, but they are very active; they manufacture various substances; they walk around; they wiggle; and they do all kinds of marvelous things—all on a very small scale. Also they store information. Consider the possibility that we too can make a thing very small which does what we want—that we can manufacture an object that maneuvers at that level. (Feynman 1960)

Nanotechnology has been identified with two primary meanings: (1) new science and technology that take advantage of nanoscale properties—that, is research, development, and commercial applications that are taking place today; and (2) using molecular machine systems to build products with atomic precision, which is an ambitious technological goal at least a decade or so away (Peterson 2004). As Merkle (1997) points out, modern-day products are made of atoms and if products and molecular machines can be built by manipulating and arranging the atoms, then there are both economic and societal benefits and costs associated with such technology. The idea of molecular manufacturing was introduced by Feynman (1960) in his talk, during which he described building using atomic precision: "if we go down far enough, all devices can be mass produced so that they are absolutely perfect copies of one another."

The idea of nanotechnology was further explored by Drexler (1981, 1992), who illustrated a design for a molecular-scale robotic arm for *positional control*. Positional control is one of the basic principles of nanotechnology (Merkle,1997). It deals with the idea that, at a microscopic scale, we should be able to position atoms and assemble them properly. Molecular manufacturing is a long-term view, but ambitious research programs in this area are already underway.

Even today, nanotechnology increasingly shows potential in commercial applications. In this context, the technology has been viewed from two major perspectives: (1) a top-down perspective, which involves starting with a larger piece of material and fabricating a nanostructure from it by removing material; and (2) a bottom-up perspective, where the focus is on working toward a better understanding of the chemical, biological, and quantum properties of atoms and molecules of all types of materials and their ability to automatically arrange themselves into more complex assemblies. The top-down approach uses traditional workshop or microfabrication methods where externally controlled tools are used to cut, mill, and shape materials, such as in circuits on microchips, into the desired shape and order (Royal Society and Royal Academy of Engineering 2004). In contrast, the bottom-up approach involves the building of structures, atom by atom or molecule by molecule, and the arrangement of smaller components into more complex assemblies by physical or chemical interactions between the units (Royal Society and Royal Academy of Engineering 2004). This principle of nanotechnology is also known as molecular self-assembly. Self-assembly is a well-established and powerful method of synthesizing complex molecular structures. Merkle (1997) refers to it as the art and science of arranging conditions so that the parts themselves spontaneously assemble into the desired structure. He provides an example of chemical synthesis where stirring compounds brings together complementary features that allow them to combine naturally into a larger part. Merkle (1997) argues that positional control and self-replication would enable the building of a staggering range of molecular structures, thus truly revolutionizing manufacturing.

The introduction of self-replicating molecular machine systems would provide economic opportunities as well as societal benefits. However, it also raises a number of concerns (Merkle 1997, 2001; Sheetza et al. 2005). One major social concern is the possibility that the self-replication could go beyond human control with adverse consequences to the world (Drexler 1986; Sheetza et al. 2005). Another major concern lies in the potential for certain groups, such as terrorists, to use the technology for deliberate abuse or for malign intent. In addition, environmental, health, and safety issues are likely to arise with developments in nanotechnology, as identified in nanotoxicology research. In order to provide a basis for informed policy decisions by citizens and governments and for the responsible development of productive nanotechnology by practitioners and industry, the Foresight Institute has developed some guidelines (2006). These guidelines are designed to address the potential positive and negative consequences of this new technology in an open and scientifically accurate manner.

13.3   NANOTECHNOLOGY R&D INVESTMENT FUNDING

The U.S. government funding for nanotechnology research has increased tremendously from US$465 million in 2001 to approximately US$1.45 billion in 2008 since the inception of the National Nanotechnology Initiative (NNI) in 2001 (National Science and Technology Council, 2007). The NNI's objective is to promote and accelerate the responsible development and application of nanotechnology to create employment, foster economic growth, and enhance national security and quality of life. On a global scale, Roco (2005) and Siegel et al. (1999) report that the worldwide investment in nanotechnology R&D has increased approximately nine-fold, from US$432 million in 1997 to about US$4.1 billion in 2005 (see Figure 13.1). A significant portion of the worldwide nanotechnology R&D investments has been by industrial countries such as the United States, Japan, and the European Union (EU); however, emerging major economic players such as China and India have also launched ambitious programs in nanotechnology R&D. Current estimates indicate that to date, about 60 countries have initiated activities in this field (Roco 2005).

At the nanoscale level, the physical, chemical, and biological properties of material may differ in fundamental and valuable ways from the properties of individual atoms, molecules, or bulk matter. R&D in nanotechnology is aimed at understanding and creating improved materials, devices, and systems that exploit these new properties (National Science and Technology Council 2007). NNI's strategic plan has funded seven categories of programs, as listed in Table 13.1. In addition, large private sector companies in automotive, aerospace, medical, and chemical industries have funded nanotechnology R&D research that is comparable to or exceeds federal levels of investment (National Science and Technology Council 2007).

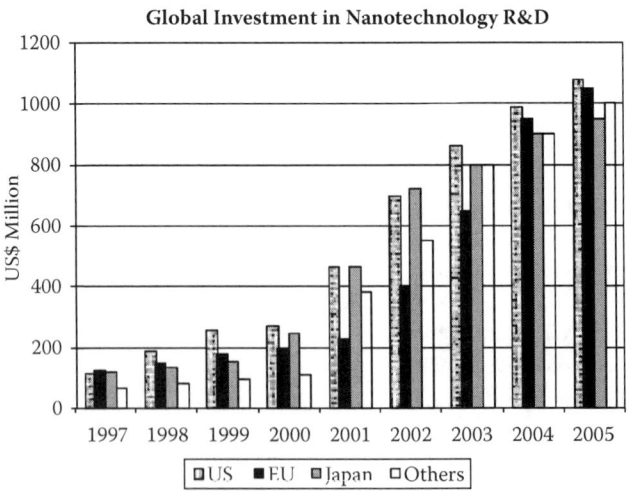

FIGURE 13.1 Worldwide strategic investments in nanotechnology (1997–2003).

TABLE 13.1

U.S. Federal Agency Investments in Nanotechnology R&D (2006–2008)*

Category	Description	$ Million		
		2006	**2007**	**2008***
Fundamental nanoscale phenomena and processes	Discovery and development of fundamental knowledge in physical, biological, and engineering sciences	455.9	436.5	491.8
Nanomaterials	Discovery, design, and synthesis of novel nanoscale and nanostructured materials with targeted properties	265.1	240.4	290.7
Nanoscale devices and systems	R&D to create novel or improve existing devices and systems using principles of nanoscience and engineering	319.6	319.1	277.4
Instrumentation research, metrology, and standards for nanotechnology	R&D related to tools required to advance nanotechnology research and commercialization	51.0	64.5	83.6
Nanomanufacturing	R&D to enable scaled-up, reliable, and cost-effective nanoscale materials, structures, devices, and systems; development of ultraminiaturized top-down processes and complex bottom-up or self-assembly processes	33.8	45.1	44
Major research facilities and instrumentation acquisition	Establishment of user facilities and networks that support or enhance scientific infrastructure in nanotechnology	152.4	162.4	159.8
Societal dimensions	Research and activities related to the assessment of environmental, health and safety costs, benefits, and risks	73.5	85.9	97.5
Total		1351.2	1353.9	1444.8

Budgeted.
Source: National Science and Technology Council (2004, 2007).

13.4 VALUING R&D AS REAL OPTIONS

R&D projects have characteristics that require them to be valued as real options. Typical characteristics include large up-front irreversible investments costs, uncertainty with respect to R&D outcomes, unavailability of immediate payoffs from commercialization of products and applications, uncertainty with respect to benefits (tangible and intangible) that are contingent on successful research, the existence of multiple sources of technology-related uncertainty, and multiple sources of uncertainty pertaining to revenue and cost parameters that make up the project cash flows. When undertaking irreversible investments such as R&D with technological and market uncertainties, firms and federal funding agencies should undertake them in an incremental manner rather than as one-time, large investments. Doing so recognizes that uncertainty creates opportunities to exploit the upside potential of investment projects by undertaking investments in stages with uncertainty resolution occurring over time and/or by active experimentation.

With the development of real options analysis as an alternative valuation tool to traditional net present value (NPV) in financial economics, several researchers have proposed valuing R&D as real options. The idea of valuing R&D as real options was proposed in the seminal work of Myers (1977, 1984) and Kester (1984). Since then, several researchers have attempted to value R&D as real options. Morris et al. (1991) applies decision tree analysis to demonstrate that given multiple R&D projects with the same expected NPVs but different levels of risk and uncertainty, under capital rationing, the riskier R&D project has more value. Faulkner (1996) provides a survey of the real options approach to valuing R&D investments. Newton and Pearson (1994), Lint and Pennings (1998), Herath and Park (1999), Kellogg and Charnes (2000), Angelis (2000), and Benninga and Tolkowsky (2000) model R&D investments as *simple real options* using the binomial and the Black and Scholes (1973) model. Pennings and Lint (1997) develop a jump diffusion model for valuing R&D. Huchzermeier and Loch (2001) incorporate budget, product performance, market requirements, and project schedule uncertainty in addition to the market payoff uncertainty in the evaluation of R&D real options.

Several researchers have emphasized that R&D should be modeled as compound options since these options are more appropriately undertaken as multi-staged type investments (Copeland and Keenan 1998; Cortazar and Schwartz 1993; Damodarana 2000). A typical R&D problem involves two basic phases, a research and development phase and a commercialization phase. Ideally, firms should make investments in R&D and commercialization in stages in each of these two phases. For example, research and development should be undertaken incrementally, thereby managing technological risk by resolving uncertainty. Similarly, commercialization should be undertaken in stages with an initial green-field entry and subsequent expansions into different markets depending on the growth of product demand.

A compound option is an option on an option (i.e., its value depends on other options). There are two types of compound options: simultaneous compound options (Geske 1979) and sequential compound options (Copeland and Antikarov 2001). In simultaneous compound options, the value of the option depends on another option, and both options are alive at the same time. A sequential compound option is a staged option. In it, the value of the upstream option depends on the downstream option value, although the order of exercise is opposite. Recently, real options researchers have attempted to model R&D and multistage capital investments as sequential compound real options (Copeland and Antikarov 2001; Herath and Park 2002). In Benaroch et al. (2006), the authors apply the R&D sequential compound real options model proposed by Herath and Park (2002) to a platform-type information technology project.

The real options-based R&D valuation models described above have only attempted to model growth (expansion) options, abandonment options, and wait-and-see (delay) options based on the standard real options framework where the exercise decision does not directly depend on the actions of competitors or collaborators. The other approaches explicitly consider (1) firm and competitors' strategic interaction in a real options setting, and (2) the firm and collaborator (partner) joint actions. There are several research attempts in this direction. Smit and Trigeorgis (2001) integrate real options and a game theoretic framework to model strategic interactions in R&D as a two-person game where the growth option value depends on exogenous competitive reactions. The basic problem setting is that a firm invests in R&D to develop a more cost-efficient production process that lowers production costs and then commercializes the product. Savva and Scholtes (2005) distinguish between cooperative options, which are exercised jointly to maximize the total deal value, and competitive options, which are exercised unilaterally to increase the payoff for individual parties. They investigate generic partnership deals under economic uncertainty with downstream flexibility. The collaborative approach falls within the boundaries of research in risk-sharing contracts.

13.4.1 NANOTECHNOLOGY R&D: UNIQUE FEATURES AND IMPLICATIONS FOR REAL OPTIONS VALUATION

Nanotechnology research and development influence several areas including manufacturing, information management, environment, and people. In the modeling of nanotechnology R&D real options, one of the main challenges is to quantify nanotechnology capabilities. R&D nanotechnology capabilities can be both tangible and intangible. For example, while firms may assess cost savings relatively easily due to nanoscale material, it will be quite a challenge for them to place a value on the benefit of radically improved drug formulas. Up-front nanotechnology R&D investment costs are relatively easy to quantify, but the unique features of nanotechnology also increase the challenges and risks of such investments. Table 13.2 lists some of the potential capabilities and benefits of nanotechnology as well as the potential costs/risks.

Nanotechnology research is interdisciplinary and its applications span several industries. Thus, collaborative research, including international collaborations, better ensures the success of nanotechnology R&D initiatives by a group or research networks investment. The interdependencies arising in nanotechnology R&D networks and the challenges and risks of nanotechnology add complexity to the valuation of R&D investments and risk management.

There are several other key features of nanotechnology R&D that should be factored into valuation. Nanotechnology research can be any one or any combination of the following: basic or knowledge-inspired research, use-inspired research, and technology development. Three nanotechnology research and development phases can be identified: fundamental discovery, technological applications, and commercialization. Nanotechnology R&D also has potential implications for health and environment, and these have to be considered in the evaluation of nanotechnology R&D investments.

TABLE 13.2

Nanotechnology Potential Benefits and Risks

Potential Capabilities/Benefits	Potential Costs/Risks
• Cost saving due to nanoscale materials and self-replication	• Capital intensive technology
• Clean and highly efficient manufacturing	• Complex R&D infrastructures
• Cheap and powerful energy generation	• Long payback time
• Greater production volume and productivity	• High degree of technology and market risks and uncertainty
• Greater information storage capability	• Environmental issues and possible regulation
• Efficient communication capacities	• Health risks
• Super computing capability	• Societal issues
• Environmentally benign resources	
• Greater agricultural productivity	
• Interactive "smart" foods and appliances	
• Improved drugs, diagnostics, and organ placement	

13.4.2 NANOTECHNOLOGY: COOPERATIVE, COMPETITIVE, OR STANDARD REAL OPTIONS?

The multidisciplinary nature of nanotechnology R&D research and its challenges, risks, and uncertainties complicate the valuation of such investments. Nanotechnology R&D investments are characterized by the irreversibility and uncertainty of technology, future benefits, and costs. As new information arrives, the uncertainties about technology, benefits, and costs are gradually resolved and firms have the flexibility to alter their R&D investment strategies. Many aspects of nanotechnology research are daunting. Rather than attempting to approach the research in a single investment, firms should instead approach nanotechnology research and development in an incremental fashion. This way, firms would be able to learn from the successes and failures of initial fundamental research and build a series of intermediate research programs, each contributing to the next research program and, eventually, commercialization. The incremental approach also should be adopted in the application development and commercialization phases. Thus, the valuation of nanotechnology R&D can be regarded as the value of a *complex compound option* in a series of follow-on investments.

Due to the fundamental nature of nanotechnology research, instead of two basic phases in a typical R&D project, namely, a R&D phase and a commercialization phase, a typical nanotechnology project would have three phases: an exploratory research phase, a R&D phase, and a commercialization phase, all with distinct stages as shown in Figure 13.2. Nanotechnology R&D would involve the resolution of multiple sources of uncertainty. These include the resolution of technical uncertainty regarding the outcome of exploratory research, the resolution of technical uncertainty regarding postexploratory R&D, and a market uncertainty resolution with regard to demand for nanotechnology applications.

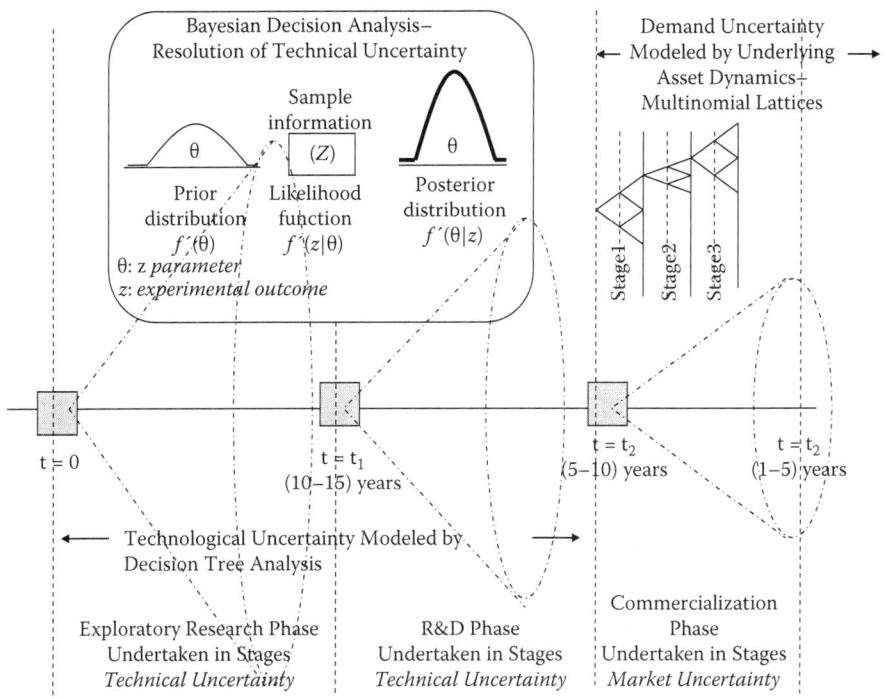

FIGURE 13.2 Nanotechnology R&D phases.

13.4.2.1 Exploratory Research

Nanotechnology involves the measuring and manipulation of matter at the nanoscale level, and, hence, much of the initial research is fundamental in nature. Exploratory research is often conducted when problems are not clearly defined; the basic premise is to discover. Whether or not the research project achieves its initial goal is not important because fundamental research can yield results in unexpected ways. There is no way to determine the exact time span involved in exploratory research, and thus we have estimated it to be at least 10–15 years. Furthermore, the outcomes are highly uncertain—shown by the cone of uncertainty in Figure 13.2. In this phase, technological uncertainty should be modeled using decision tree analysis. One component of NNI R&D funding is primarily aimed at exploratory research that could lead to the discovery and development of novel ideas through the investigation of fundamental nanoscale phenomena, properties, processes, structures, and architecture, as well as the development of experimental and simulation tools (National Science and Technology Council 2004).

13.4.2.2 Postexploratory R&D

In the postexploratory R&D phase, the problems investigated are clearly defined. This phase would utilize the knowledge from the exploratory phase and focus on incubating new concepts for industrial applications. Again, the time span of R&D success is not known (perhaps 5–10 years), but the uncertainty has been somewhat resolved, thus the cone of uncertainty is smaller than it is in the exploratory phase. Here again, technological uncertainty could be modeled using decision tree analysis.

13.4.2.3   Commercialization

The commercialization phase involves the marketing of nanotechnology applications in existing products and the development of new products, processes, and applications. There are a multitude of uncertainties in commercialization, including the timing of entry into markets, the scale and scope of entry, and the demand among others. Additionally, the nature of nanotechnology itself adds another risk dimension that is unique and significant—that is, uncertainty concerning the safety of humans and the environment. As we have shown in Figure 13.2, market uncertainty should be modeled by lattices (binomial or multinomial). Lattice models provide more latitude in capturing such complexities as multiple underlying assets that are correlated, uncertain exercise prices, optimal timing, and varying asset volatilities than continuous time models.

A multitude of real options may arise depending on the type of nanotechnology research, which industry it pertains to, the type and life cycles of products, and the products and applications available for commercialization, among others. These real options are also further complicated by how the firms and governments would make investment decisions. For example, they could decide to invest in nanotechnology R&D independently, form R&D networks, invest jointly through R&D partnerships, or a combination of these. As we have discussed, due to these heterogeneous features, corporate decision makers have to develop situation specific models to argue the case for investment in nanotechnology R&D. Also, since there will be budget considerations, environmental and health constraints, and portfolio effects, nanotechnology R&D models may have to consider these effects and constraints as well. We thus discuss the generic types of real options that may arise in any of the three phases, exploratory, R&D, and commercialization. Depending on the problem instances, analysts can choose the real options that are applicable to developing an integrated compound R&D real options model.

The real options literature has identified several basic types of real options, namely:

1. *Option to expand*: If the initial investment turns out well, then firms can exercise the option to expand operations.
2. *Option to switch*: In some situations, firms may have the option to switch use between inputs.
3. *Option to delay investment*: Sometimes, delaying an investment may turn out to be more profitable, as when waiting adds value through the resolution of uncertainty.
4. *Option to abandon*: Firms can make a small initial trial investment and exercise the abandonment option if events do not go as planned.

In Figure 13.3, we position the four basic real options in the context of nanotechnology.

In addition to these four basic options, we identify three more real options that arise in nanotechnology R&D.

Option to Expand	Option to Switch
• Supplements the bottom-up perspective • Allows learning over time scope by scope • Incremental adoption • May face compatibility issues with exiting technologies for simultaneous adoption	• Allows resource reallocation • Allows extension to other applications • May face flexibility issues because of possible incompatibility with current technologies
Option to Delay Investment	**Option to Abandon**
• Long-term value could be substantial and significant affecting multiple industries • Immediate adoption has substantial risk • Waiting could mitigate the high uncertainty risk associated with immediate adoption • Waiting could add value if technology is harmful to society and environment	• Nanotechnology is harmful to humans and environment • Market demand diminished • May face difficulty in assessing salvage value

FIGURE 13.3 Basic option scenarios for nanotechnology R&D.

5. *Cooperative options*: Due to the fundamental nature of nanotechnology research, there is an immense uncertainty associated with the outcome of R&D. Rather than investing large amounts in basic research as a go-it-alone investment strategy, firms can enter into R&D partnerships or networks that minimize individual firm risks. The opportunity of collaborating allows for the sharing of nanotechnology R&D risks. Savva and Scholtes (2005) model cooperative options as partnership deals that have future flexibility. In partnership deals, the exercise decisions are taken jointly with a view to maximizing the total value of the deal. Corporate planners who are making the case for R&D investments can also draw upon financial risk-sharing and insurance literature when modeling cooperative options.

6. *Competitive options*: Firms should also have the flexibility of going alone rather than collaborating, especially in the postexploratory commercialization phases. If a firm decides to invest in nanotechnology R&D independently, then it is competing in R&D. Here the strategic interactions of players are important and real options could be integrated with game theory (Smit and Trigeorgis 2001). Competitive action may also prompt preemptive investments in nanotechnology R&D and/or preemptive market entry to obtain a first-mover advantage.

7. *Learning options*: The sequential investments in R&D phases, as well as in the commercialization phase, allow for learning. In modeling learning options, one needs to distinguish between learning curve effects that affect production cost functions and learning through active experimentation (or

Bayesian learning; (Beirman and Rao 1978; Herath and Park 2001). Both are applicable in valuing nanotechnology R&D.

Such competitive interactions or collaborative actions add complexity into standard real options, such as those in R&D commercialization. In valuing nanotechnology R&D, corporate planners should choose which of the above-listed real options should be incorporated in the compound R&D real option. The resulting model would be an integrated compound R&D real options model. In addition to economic issues, R&D decisions on new technologies have to take into account a number of social, environmental, and health issues. These factors require new approaches that incorporate additional criteria along with quantitative values obtained from an integrated compound real options model. As nanotechnology R&D investment decisions are made, a comparative assessment of options is needed at each stage.

13.5   NANOTECHNOLOGY R&D RISK MANAGEMENT

Nanotechnology has widely been described as a general-purpose or transformative technology with the potential to bring considerable benefits to society (Maynard 2006; Rickerby and Morrison 2007; Walsh et al. 2008). This calls for significant efforts on the part of R&D management not only to prevent adverse consequences at an early stage in the development cycle but also to expand beyond traditional risk management (Maynard 2006). The lack of well-defined scientific knowledge regarding the impact of nanotechnology requires new risk management approaches beyond simple cost-benefit analysis (Renn and Roco 2006). Potential nanotechnology applications are subject to the jurisdictions of multiple regulatory programs; however, existing regulatory policies may not be adequate to address the potential risks to workers, consumers, the public, and the environment (Walsh et al. 2008). Because nanotechnology raises issues that are more complex and far-reaching than those of many other innovations, there is a need for a system of risk governance that is global, is coordinated, and involves the right mix of voluntary corporate leadership; coordinated private, academic, and government research; and informed regulation (International Risk Governance Council 2007; Krupp and Holliday 2005).

In order to manage nanotechnology risks, the International Risk Governance Council (IRGC) provides a policy brief that identifies key areas where relevant stakeholders could contribute to improved risk and benefit governance in a coordinated fashion (IRGC 2007). This policy brief provides risk appraisals related to health, environment, manufacturing, education gaps, public perceptions, ethics, politics and security, and transboundary risks. One of the key observations is that the policy and regulatory environment is far behind the pace of the technological progress, and thus the current systems of risk governance are incomplete and fragmented (IRGC 2007; Renn and Roco 2006). In response to these risk governance gaps, the IRGC has made recommendations in the following five major categories:

1. Improve the knowledge base.
2. Strengthen risk management structures and processes.

3. Promote stakeholder communication and participation.
4. Ensure social benefits and acceptance.
5. Collaborate between stakeholders and nations.

Considering the high uncertainty surrounding nanotechnology R&D as well as the limitations of existing regulatory tools and policies, Walsh et al. (2008) suggest three distinct initiatives that should be addressed urgently:

1. A major increase in nanomaterial risk research in order to effectively assess nanotechnology's health and environmental implications
2. The rapid development and implementation of voluntary standards of care, which encompass a framework and processes to identify and manage nano-materials' risks across their full product life cycles, including worker safety, manufacturing releases, product use, and product disposal
3. The development of an adequate regulatory framework for nanomaterial risk management in order to secure sustained public confidence and sup-port for nanotechnology

Given these recommendations, it is obvious that nanotechnology R&D risk management requires broad collaboration and inclusive stakeholder engagement, including labor and community groups, health and environmental organizations, consumer advocates, large and small businesses, and the academic community (Krupp and Holliday 2005; Walsh et al. 2008). Some progress toward this objec-tive is already being seen in the partnership between Environmental Defense and DuPont (2007). The objective of this partnership is to develop a nano risk framework across a product's life cycle and provide guidance for organizations in developing applications and on effective risk evaluations and risk management decisions. In particular, this framework addresses the following specific needs:

1. It allows users to address areas of incomplete or uncertain informa-tion by using reasonable assumptions and appropriate risk management practices.
2. It describes a system designed to guide information generation and update assumptions, decisions, and practices for dealing with new information as it becomes available.
3. It offers guidance on how to communicate information and decisions to stakeholders.

Such a framework facilitates the generation and communication of information regarding environmental, health, and safety exposures, and regulatory issues that could help organizations assess risks more effectively, develop risk-preventive strate-gies, and reduce losses. In addition, it provides relevant information for the insurance industry to facilitate risk management services, as well as for the investment com-munity to build confidence in long-term investments.

Furthermore, nanotechnology R&D risk management can be examined from a "technology assessment" perspective, which aims to provide knowledge and orientation for acting and decision making concerning technology and its implementation in society (Bechmann et al. 2007; Decker and Ladikas 2004; Fleischer and Grunwald 2008). In this context, technology assessment provides advice to decision makers on possible side effects and risks, as well as the potentials and benefits of nanotechnology at early stages of development and optimal strategies for harvesting opportunities (Fleischer and Grunwald 2008; von Schomberg 2005).

13.6 CONCLUSIONS AND DIRECTIONS FOR FUTURE RESEARCH

In this chapter, we have explored a range of issues associated with the application of real options theory for nanotechnology R&D investment project valuation. We have described nanotechnology and discussed the opportunities, uncertainty, and risks surrounding nanotechnology R&D projects. Our goals have been to explore how real options theory can be used as an effective valuation method and to provide a broad framework for short-term and long-term risk management of nanotechnology R&D projects.

The multidisciplinary nature of nanotechnology and its potential impact on a wide number of industries offer a broad scope for numerous opportunities. As a result, funding for nanotechnology research has increased tremendously in many industrial nations. Paralleling this trend has been an increase in awareness of nanotechnology and its commercial feasibility and risks. Our review of the opportunities and challenges shows the unique features of nanotechnology R&D projects and the complexities in the R&D project phases that factor into valuation. We have identified different types of real options that may arise in the different phases of a nanotechnology R&D project.

Dynamic R&D risk management would require modeling real options and new risk frameworks that are comprehensive and effective in addressing the health and environmental issues and risks that span several industries and sectors. In other words, there is a need for collaboration in order to understand the complexities, educate the public, and address the environmental, health, economic, and social risks. This calls for inclusive stakeholder engagement and the involvement of government, policy makers, academics, the private sector, and communities.

When considering the widespread and pervasive impact of nanotechnology on society, it becomes apparent that much remains to be researched in nanotechnology R&D. Future research can address several issues regarding policy making, as well as modeling and analysis angles. These include, but are not limited to, the development of complex compound real options with Bayesian learning, the development of analytical models to value nanotechnology R&D portfolios, constrained R&D portfolio optimization models, extreme event modeling, and the development of a set of comprehensive risk metrics that includes both quantitative and qualitative factors.

REFERENCES

Angelis, L. 2000. Capturing the option value of R&D. *Research Technology Management* (July–August): 31–34.

Bechmann, G., Decker, M., Fiedeler, U., and Krings, B.-J. 2007. Technology assessment in a complex world. *International Journal of Foresight and Innovation Policy* 3(1): 6–27.

Beirman, H., Jr., and Rao, V. 1978. Investment decisions with sampling. *Financial Management* 7(7): 19–24.

Benaroch, M., and Kauffman, R. J. 1999. A case for using real options pricing analysis to evaluate information technology project investments. *Information Systems Research* 10(1): 70–86.

Benaroch, M., Shah, S., and Jeffery, M. 2006. On the valuation of multistage information technology investments embedding nested real options. *Journal of Management Information Systems* 23(1): 239–261.

Benninga, S., and Tolkowsky, E. 2002. Real options: An introduction and an application to R&D valuation. *The Engineering Economist* 47(2): 151–168.

Bhat, J. S. A. 2005. Concerns of new technology based industries: The case of nanotechnology. *Technovation* 25(5): 457–462.

Black, F., and Scholes, M. 1973. The pricing of options and corporate liabilities. *Journal of Political Economy* 81(3): 637–654.

Copeland, T. E., and Antikarov, V. 2001. *Real options: A practitioner's guide*, 1st ed. New York: Texere.

Copeland, T. E., and Keenan, P. T. 1998. Making real options real. *The McKinsey Quarterly* (3): 128–141.

Cortazar, G., and Schwartz, E. S. 1993. A compound option model of production and intermediate inventories. *Journal of Business* 66(4): 517–540.

Damodarana, A. 2000. The promise of real options. *Journal of Applied Corporate Finance* 13(9): 29–44.

Decker, M., and Ladikas, M. 2004. *Bridges between science, society, and policy: Technology assessment—methods and impacts*. Berlin: Springer.

Drexler, K. E. 1981. Molecular engineering: An approach to the development of general capabilities for molecular manipulation. *Proceedings of the National Academy of Science* 78(9): 5275–5278.

Drexler, K. E. 1986. *Engines of creation: The coming era of nanotechnology*. New York: Anchor-Doubleday.

Drexler, K .E. 1992. *Nanosystems: Molecular machinery, manufacturing, and computation*. New York: John Wiley.

Environmental Defense and DuPont. 2007. Nano risk framework. http://www.NanoRiskFramework.com.

Faulkner, T. W. 1996. Applying "options-thinking" to R&D valuation. *Research: Technology Management* 39(3): 50–56.

Feynman, R. 1960. There's plenty of room at the bottom: An invitation to enter a new field of physics. *Engineering and Science Magazine* 23(5).

Fleischer, T., and Grunwald, A. 2008. Making nanotechnology developments sustainable: A role for technology assessment? *Journal of Cleaner Production* 16(8–9): 889–898.

Foresight Institute. 2006. Foresight guidelines for responsible nanotechnology development, draft version 6, April. http://www.foresight.org/guidelines.

Geske, R. 1979. The valuation of compound options. *Journal of Financial Economics* 7(1): 63–81.

Guz, I. A, Rodger, A. A., Guz, A. N., and Rushchitsky, J. J. 2007. Developing the mechanical models for nanomaterials. *Composites Part A: Applied Science and Manufacturing* 38(4): 1234–1250.

Herath, H. S. B., and Park, C. S. 1999. Economic analysis of R&D projects: An options approach. *The Engineering Economist* 44(1): 1–35.

Herath, H., and Park, C. 2001. Real option valuation and its relationship to Bayesian decision-making methods. *The Engineering Economist* 46(1): 1–32.

Herath, H. S. B., and Park, C. S. 2002. Multi-stage capital investment opportunities as compound real options. *The Engineering Economist* 47(1): 1–27.

Huchzermeier, A., and Loch, C. H. 2001. Project management under risk: Using the real options approach to evaluate flexibility in R&D. *Management Science* 47(1): 85–101.

International Risk Governance Council (IRGC). 2007. *Policy brief: Nanotechnology risk governance: Recommendations for a global, coordinated approach to the governance of potential risks.* Geneva: IRGC.

Kellogg, D., and Charnes, J. M. 2000. Real-options valuation for a biotechnology company. *Financial Analysts Journal,* 56(3): 76–84.

Kester, W. C. 1984. Today's options for tomorrow's growth. *Harvard Business Review,* (March–April): 153–160.

Krupp, F., and Holliday, C. 2005. Let's get nanotech right. *Wall Street Journal,* June 14.

Kumar, R. L. 2002. Managing risks in IT projects: An options perspective. *Information & Management* 40(1): 63–74.

Lint, O., and Pennings, E. 1998. R&D as an option on market introduction. *R&D Management* 28(4): 279–287.

Maynard, A. D. 2006. Nanotechnology: Assessing the risks. *NanoToday* 1(2): 22–33.

Merkle, R. C. 1997. It's a small, small, small, small world. *MIT's Technology Review* 100(2): 25–32.

Merkle, R. C. 2001. Nanotechnology: What will it mean? *IEEE Spectrum* 38(1): January 19–21.

Morris, P. A., Teisberg, E. O., and Kolbe, A. L. 1991. When choosing R& D go with long shots. *Research: Technology Management:* 34(1): 35–40.

Myers, S. C. 1977. Determinants of corporate borrowing. *Journal of Financial Economics* 5(2): 147–176.

Myers, S. C. 1984. Financial theory and financial strategy. *Interfaces* 14(January–February): 126–137.

National Science and Technology Council. 2004. *The National Nanotechnology Initiative: Strategic Plan,* December. Washington, DC: NSTC.

National Science and Technology Council. 2007. *The National Nanotechnology Initiative: Supplement to the President's 2008 Budget,* July. Washington, DC: NSTC.

Newton, D. P., and Pearson, A. W. 1994. Application of option pricing theory to R&D. *R&D Management* 24(1): 83–89.

Pennings, E., and Lint, O. 1997. The option value of advanced R & D. *European Journal of Operational Research* 103(1): 83–94.

Perlitz, M., Peske, T., and Schrank, R. 1999. Real options valuations: The new frontier in R&D project valuation? *R&D Management* 29(3): 255–269.

Peterson, C. L. 2004. Nanotechnology: From Feynman to the grand challenges of molecular manufacturing. *IEEE Technology and Society Magazine* 23(4): 9–15.

Renn, O., and Roco, M. C. 2006. Nanotechnology and the need for risk governance. *Journal of Nanoparticle Research* 8(2): 153–191.

Rickerby, D. G., and Morrison, M. 2007. Nanotechnology and the environment: A European perspective. *Science and Technology of Advanced Materials* 8(1–2): 19–24.

Roco, M. C. 2005. International perspective on government nanotechnology funding in 2005. *Journal of Nanoparticle Research* 7(6): 707–712.

Royal Society and Royal Academy of Engineering. 2004. Nanoscience and nanotechnologies: Opportunities and uncertainties. http://www.nanotec.org.uk/finalReport.htm. Retrieved August 28, 2004.

Savva, N., and Scholtes, S. 2005. Real options in partnership deals: The perspective of cooperative game theory. Discussion paper. Cambridge: Judge Business School, University of Cambridge.

von Schomberg, R. 2005. The precautionary principle and its normative challenges. In *The precautionary principle and public policy decision-making,* ed. E. Fisher, J. Jones, and R. von Schomberg. Cheltenham, UK: Edward Elgar. http://www.eng.cam.ac.uk/~ss248/publications/ROP15Feb2006.pdf.

Sheetza, T., Vidalb, J., Thomas, P. D., and Lozanoa, K. 2005. Nanotechnology: Awareness and societal concerns. *Technology in Society* 27(3): 329–345.

Siegel, R. W., Hu, E., and Roco, M. C., eds. 1999. *Nanostructure science and technology, NSTC, Washington, D.C.* Boston: Kluwer Academic.

Smit, H. T. J., and Trigeorgis, L. 2001. R&D option strategies. *5th Annual Real Options Conference,* Los Angeles, June.

Walsh, S., Balbus, J. M., Denison, R., and Florini, K. 2008. Nanotechnology: Getting it right the first time. *Journal of Cleaner Production* 16:1018–1020.

14 Real Options-Based Analysis in Pharmaceutical Partnerships for Research and Development

Minjung Kim
Pennsylvania State University

CONTENTS

This chapter suggests a methodology for determining the optimal timing and partnership terms in a real options approach that provides managerial flexibility. In this model, the decision to invest in a new drug development as an aspect of a pharmaceutical company's consideration is represented as exercising a call option. A decision to sell ownership of a new drug is considered as a biotechnology company exercising a put option. Based on this structure, a model to determine the optimal timing is proposed by considering ownership ratio, synergy effect, and payment options.

14.1 PARTNERSHIPS IN RESEARCH AND DEVELOPMENT

As the complexity of scientific and technological development increases, the uncertainty surrounding a research and development (R&D) project increases; consequently, the cost of the R&D project increases astronomically. To cope with the high level of uncertainty and enormous R&D expense, the number of partnerships among companies that may complement each other has rapidly increased since the 1990s. According to Hagedoorn (2002), medical technology leads the trend toward R&D partnerships by occupying 80 percent of this domain.

Numerous studies have emphasized the importance of real options analysis for corporate managerial decision making, including McDonald and Siegel (1986), Brennan and Schwartz (1985), Dixit (1989), and Grenadier and Weiss (1997). Dixit and Pindyck (1994) mentioned that irreversible investment can be explained as a perpetual financial option, and the real options theory can provide more accurate valuation of project. The model suggested by Dixit and Pindyck (1994) showed that an investment can be exercised when a project's value is larger than or equal to total cost and that its optimal timing to invest can be determined when the option premium has the largest positive value. Clark and Rousseau (2002) provided a study of an optimal timing problem for an option to abandon. Park and Herath (2000) illustrated the value of managerial flexibility to delay a project.

A multistage investment such as a new drug development project is an especially good candidate for evaluation using real options because of its intrinsic characteristics: high uncertainty associated with project cash flows, unavailability of immediate payoffs, and multiple sources of uncertainty at every project phase. Application of real options pricing has been proposed for value in R&D projects in the works of Myers (1984), Nichols (1994), Herath and Park (1999), Herath and Park (2002), and Rogers et al. (2005). For instance, Mitchell and Hamilton (1988) applied real options to drug valuation with the price of a call option on the future commercialization of the project. Rogers et al. (2002) solved the portfolio selection decision using real options valuation in view of pharmaceutical companies.

The overall trends and factors impacting partnerships between pharmaceutical companies and biotechnology companies have had significant examination. Cohen et al. (2005), Smith (2006), and Arnold et al. (2002) all studied the advantages and disadvantages of partnerships in medical fields and elaborated the synergistic effects throughout a partnership by providing case studies. Several authors have addressed partnerships on the basis of a real options approach. Rogers et al. (2005) evaluates the R&D licensing opportunities in the aspect of pharmaceutical companies based on the real options to determine the best licensing timing, and Merck applied real options with respect to valuation of biotechnology investment (Nichols 1994). However, we rarely see papers that analyze licensing in a view of biotechnology companies because most studies in licensing alliances in the medical field are focused on only pharmaceutical companies. Also, little work has been done to understand the effect of real options in light of synergies created by licensing and the optimal timing and subsequent terms that both of companies can satisfy bilaterally as well as obtain maximum gains in terms of finance and R & D. The aim of this chapter is to fill this gap by presenting a framework that explains the interrelationship between two companies in licensing based on the real options methodology.

Viewing partnerships in the medical industry in specific, newly established pharmaceutical-biotechnological alliances is the main partnership structure to complete new drug development projects. The partnership between a big pharmaceutical company that has capital and marketing resources and a biochemical company that possesses the intellectual property of a candidate drug has become a trend in new drug development projects. While preoccupation of market share is important for big pharmaceutical companies, such positioning is beyond a small biotechnology company's capabilities to deal with the substantial time and huge investment required to commercialize a product.

To realize a mutually beneficial partnership, conditions and timing should satisfy both parties. For instance, pharmaceutical companies that are licensers do not achieve a partnership when the value of the project is less than the total licensing payment. Biotechnology companies also would agree to a partnership when the value of the transferred project is larger than the alliance payment. Moreover, since the value of the project varies as factors surrounding the market and the project change, the decisions regarding optimal conditions and timing of an alliance are crucial and need careful examination.

14.2 OPTIMAL TIMING MODEL

In many real options studies of general R&D projects, the optimal timing decision gives a project manager a threshold for either investing or delaying the project at every stage. This idea can be simply applied to licensing between pharmaceutical and biotechnology companies. Investing timing for pharmaceutical companies and intellectual property selling timing for biotechnology companies can be derived based on the optimal stopping problem. As shown in Figure 14.1, the partnership divides into the two viewpoints of the pharmaceutical companies and biotechnology companies. The consensus of a contract could be regarded as exercising a call option for pharmaceutical companies because the partnerships provide the opportunity to invest or not. Achieving the contract could be regarded as exercising a put option for biotechnology companies because licensing implies selling ownership of intellectual property in return for: licensing payment that bears all costs for further clinical and regulatory development, up-front payments, and milestone payments at the end of each successful stage in the development pipeline.

The calculation of call and put options uses a binomial approximation of the stochastic process based on discrete time. The binomial pricing approach, developed by Cox et al. (1979), not only is known as transparent and computationally efficient, but also suggests a closed form solution. It implies that investors can obtain a specified guideline if they track all procedures step by step.

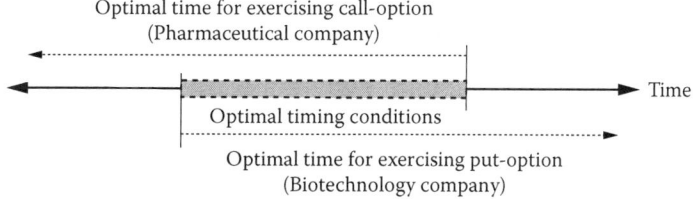

FIGURE 14.1 Relationship between a pharmaceutical company and a biotechnology company.

As a method to express the uncertainty of a new drug development project, the value of the project is assumed to follow a geometric Brownian motion (GBM) so the estimate of its value at any point in time has a lognormal distribution. The value of underlying asset value (V) can be uV with probability p, and dV with probability $(1 - p)$ in one discrete time. The value of a call/put option is obtained by discounting the expected value of the option with respect to the up and down factors (u and d), the risk-neutral probability (p), and using the risk-free rate (r_f).

For a simpler approach, one period model is observed in Figure 14.2. The main assumption is that the underlying asset (V) follows a multiplicative binomial process. The risk-free rate on a riskless asset is determined as r_f, and the strike price as K. The value of the underlying asset at the end of one period can be expressed as V_d and V_d. The option values are finally found by discounting the value of the expected output (V_d and V_d) with respect to the risk-neutral probability p to time zero.

14.2.1   Considerations for a Licensing Agreement

A licensing agreement can be made only when both sides have their own needs satisfied. Since they have different expectations for the partnership, disagreements may arise in achieving an alliance. By reflecting on their opinions in their partnership, several terms must be considered, such as ownership ratio (a), synergy effect (b), and payment options that reflect allocation of up-front payments and milestone payments at every successful stage.

14.2.1.1   Ownership Ratio

When choosing to license with a pharmaceutical company, a biotechnology company transfers a particular percentage of ownership (a) of the new drug to the pharmaceutical company, receiving in return payments for costs of further testing and

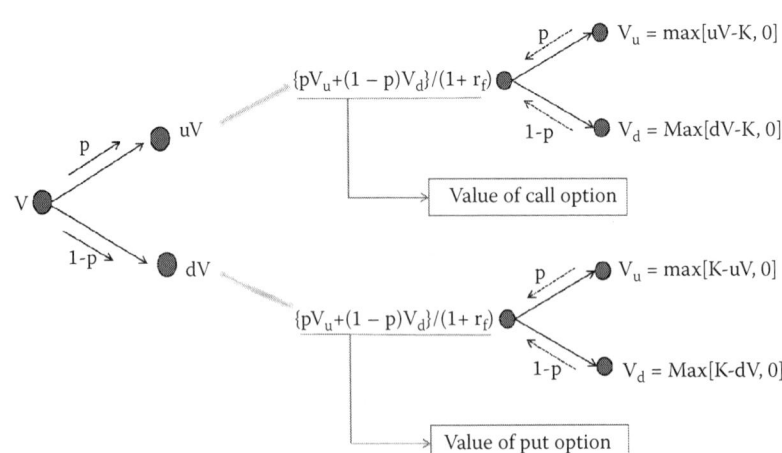

FIGURE 14.2 Binomial pricing approach.

up-front and milestone remuneration. The value of the new drug that a biotechnology company owns is multiplied by $(1 - a)$ at the original value without considering the partnership. Alternatively, the value of the new drug that a pharmaceutical company achieves is multiplied by a at an original value.

14.2.1.2 Synergy Effect

Pharmaceutical companies with advanced marketing resources and stable economic conditions may generate at least more than the value from a licensed product than smaller biotechnology companies. The synergy effect (b) is a parameter defined as the value-added contribution, from the pharmaceutical company to the value of the project. When the alliance forms, the project can have an expected value of bV_o at time $t = 0$.

14.2.1.3 Payment Options

Pharmaceutical companies and biotechnology companies have opposite prefer-ences for license payment options. The payment options should be considered according to ownership ratio when establishing the optimal timing of an alliance. Pharmaceutical companies originally take responsibility for the testing expenses as well as up-front payments and milestone payments. Up-front payments could be regarded as the first milestone payment because payment occurs when the alli-ance is attained for convenience. When considering an up-front payment as one of the milestone payments, the fraction (Y_S) of the total licensing payments at the developmental stage, S, made to a biotechnology company is a parameter that reflects strategies that each company pursues. For instance, a pharmaceutical com-pany wants a smaller up-front payment and larger milestone payments in later stages because it wants to avoid taking risks during the new drug's development process. Biotechnology companies, however, prefer to attain a license agreement with larger up-front payments and smaller milestone payments in later stages because they desire more income because of the risk that the project may fail during development. Within these constraints, payment policies can be largely classified as follows:

1. *Increasing ratio payment policy*: The payment ratio (Y_S) increases over time. It is the payment option that the large pharmaceutical companies seek.
2. *Constant ratio payment policy*: The payment ratio is always the same. The expected uncertainties are almost equally distributed.
3. *Decreasing ratio payment policy*: The payment ratio decreases over time. Biotechnology companies that chase rewards as early as possible prefer this payment option.

In detail, pharmaceutical companies want to obtain more ownership but later part-nerships and increasing ratio payment options. And biotechnology companies want to assign less ownership to pharmaceutical companies but embrace earlier partner-ship and decreasing payment ratio options. These conflicting aspirations may reveal some link among ownership ratio (a), contract timing (M), and payment options. Table 14.1 explains that a greater ownership stake requires earlier partnership and a

TABLE 14.1

Trade-Off between Money Policy and Optimal Timing versus Ownership

Contract Timing	Payment Option	Stage 1	Stage 2	Stage 3	Stage 4	Stage 5	Ownership (a)
1 (Preclinical)	Increasing ratio	0.3	0.25	0.2	0.15	0.1	0.95
	Constant ratio	0.2	0.2	0.2	0.2	0.2	0.915
	Decreasing ratio	0.1	0.15	0.2	0.25	0.3	0.880
	Increasing ratio	0	0.4	0.3	0.2	0.1	0.880
2 (Clinical I)	Constant ratio	0	0.25	0.25	0.25	0.25	0.845
	Decreasing ratio	0	0.1	0.2	0.3	0.4	0.810
	Increasing ratio	0	0	0.44	0.34	0.22	0.810
3 (Clinical II)	Constant ratio	0	0	0.33	0.34	0.33	0.775
	Decreasing ratio	0	0	0.22	0.34	0.44	0.740
	Increasing ratio	0	0	0	0.4	0.6	0.740
4 (Clinical III)	Constant ratio	0	0	0	0.5	0.5	0.705
	Decreasing ratio	0	0	0	0.6	0.4	0.670
FDA filing	Constant ratio	0	0	0	0	0	0.670

decreasing ratio money policy. Conversely, a later partnership implies smaller ownership and an increasing ratio payment option.

14.3 OPTIMAL INVESTMENT TIMING FOR BIOTECHNOLOGY COMPANIES

Biotechnology companies fear the effects of technical and market risks for a new drug development project because the staged failure comes from the abandonment of a project without receipt of any intermittent profit. Licensing can be a smart decision because it can help reduce fear of losses by guaranteeing secure money $(C \cdot Y_S)$ received from a pharmaceutical company at the end of each staged success in return for a portion of project ownership $(a \cdot V_{t=L})$ when an alliance is attained at time $t = L$. Specifically, licensing for biotechnology companies indicates a decision to sell a project $(V_{t=L})$ with staged testing cost (I_s), licensing payment $(C \cdot Y_S)$ at every stage (s), and $a \cdot b \cdot V_{t=10-L}$ when the product is commercialized.

Viewing this idea as a real option, a contract for biotechnology companies can be regarded as exercising a put option. A partnership can result only when a put option premium is positive, and optimal timing is obtained by maximizing a put option premium. When a project enters the phase of preclinical testing at $t = 0$ $(L = 0)$, the value of a project available at the end of preclinical testing $(t = 2)$ could be as follows:

- Project value at the end of Preclinical Phase: $V_{t-2} = V_0 u^{N-i} d^i$ $(i = 0, 1, 2, \ldots, N)$

Figure 14.3 illustrates the binomial tree when N (number of available scenarios) = 4 at the beginning of the current stage. The lattice steps are $i = \{0, 1, 2, \ldots, N, N+1,$

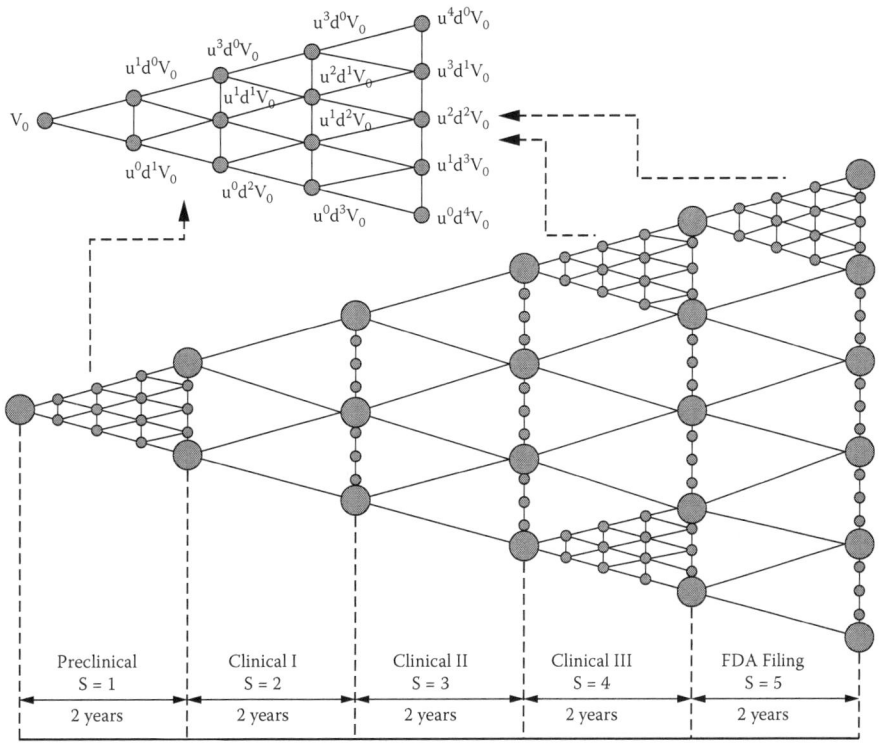

FIGURE 14.3 Binomial pricing tree showing product at end of preclinical testing (N = 4).

$N + 2, \ldots, 5 \cdot N\}$, which are shorthand as $i = \{0, 1, 2, 5 \cdot N\}$. The project values at the end of Clinical Phase I testing ($t = 4$), Phase II testing ($t = 6$), Phase III testing ($t = 8$), and FDA approvals ($t = 10$) can be calculated as follows:

- Project value at the end of Clinical Phase I: $V_{t=4} = V_0 u^{2N-i} d^i$ ($i = 0, 1, 2, \ldots, 2 \cdot N$)
- Project value at the end of Clinical Phase II: $V_{t=6} = V_0 u^{3N-i} d^i$ ($i = 0, 1, 2, \ldots, 3 \cdot N$)
- Project value at the end of Clinical Phase III: $V_{t=8} = V_0 u^{4N-i} d^i$ ($i = 0, 1, 2, \ldots, 4 \cdot N$)
- Project value at the end of FDA approval: $V_{t=10} = V_0 u^{5N-i} d^i$ ($i = 0, 1, 2, \ldots, 5 \cdot N$)

After enumerating the values for all possible scenarios in an overall framework, option value can be calculated: starting with terminal nodes, the nodes move backward to the first node throughout the intermittent nodes by backward induction. Basic considerations when calculating gains from exercising a put option are all total amounts in an alliance deal (up-front payment plus milestone payment), C; cost for all the testing and FDA registration to stage s, I_s; and probability of technical success in its stage s, Z_s. In detail, the terminal nodes are calculated through the maximization between (1) executing the option and (2) allowing the option to expire, making it worthless. If benefits of execution exceed costs, a put option will be exercised. Otherwise, the project will be abandoned. Terminal node:

$$P_{t=10}^{i_5} = \max[(CY_5 + I_5 + (1-a)bV_0 u^{5N-m}d^m)Z_5 - V_0 u^{5N-m}d^m Z_5, \ 0]$$

$$= \max[(CY_5 + I_5 + (b-ab-1)bV_0 u^{5N-m}d^m)Z_5, \ 0]$$

An additional aspect requires consideration when a put option is exercised: technical risk. An assumption necessary for rendering a put option evaluation valid is the absence of an obstruction technical failure. Even one technical failure in a project before a commercialization stage renders a project unviable. It implies that put option pricing for a project having a barrier of technical failure renders the pricing meaningless because the value of the project may ultimately be zero.

The next concern in valuation is the calculation of intermediate nodes. Intermediate nodes are calculated using a risk-neutral probability analysis until they meet the decision point where decisions are made regarding whether to exercise or abandon the option. Calculation of the decision point occurs similarly to that of the terminal node.

- Intermediate decision point:

$$P_{t=8}^{i_4=m} = \max\left(CY_4 + I_4 - \sum_{j=m}^{N+m} \binom{N}{j} p^{N-j}(1-p)^j p_{t=8}^{i_4=j} Z_3, \ 0\right)$$

$$(m = 0, 1, 2, \ldots, 4 \cdot N)$$

$$P_{t=6}^{i_3=m} = \max\left(CY_3 + I_3 - \sum_{j=m}^{N+m} \binom{N}{j} p^{N-j}(1-p)^j p_{t=8}^{i_4=j} Z_3, \ 0\right)$$

$$(m = 0, 1, 2, \ldots, 3 \cdot N)$$

$$P_{t=4}^{i_2=m} = \max\left(CY_2 + I_2 - \sum_{j=m}^{N+m} \binom{N}{j} p^{N-j}(1-p)^j p_{t=6}^{i_3=j} Z_2, \ 0\right)$$

$$(m = 0, 1, 2, \ldots, 2 \cdot N)$$

$$P_{t=2}^{i_1=m} = \max\left(CY_1 + I_1 - \sum_{j=m}^{N+m} \binom{N}{j} p^{N-j}(1-p)^j p_{t=4}^{i_2=j} Z_1, \ 0\right)$$

$$(m = 0, 1, 2, \ldots, N)$$

- Put option premium: $P_{t=0}^{i_0} = \sum_{j=0}^{N} \binom{N}{j} p^{N-j}(1-p)^j p_{t=2}^{i_1=j}$

14.4 OPTIMAL INVESTMENT TIMING FOR PHARMACEUTICAL COMPANIES

Pharmaceutical companies, before making decisions about partnerships with bio-technology companies, will compare the returns expected by acquiring a new drug development project and the licensing payment. Based on a real options framework, a licensing agreement can be regarded as exercising a call option. The initial value of the asset (V_o) is determined as the value of a project when a product is commercialized $(V_{t=10})$, multiplied by ownership ratio (a), and multiplied by the synergy effect (b). And the strike price is determined as all the expenses for the remaining testing $(I_S,$ an up-front payment, and milestone payments $C \cdot Y_S)$. Throughout a binomial pricing approach, used by a put option of biotechnology companies, the call option premium is calculated and its investment can be accomplished only when a call option premium is positive. The optimal investment rule for pharmaceutical companies can be obtained when the value of a call option premium is maximized.

The following is a case in which a partnership is attained at the preclinical testing stage. The representation shows the processes that result in a call option premium. The first step is to create a lattice considering the upward effect and downward effect at every discrete point in time.

$$V_{t=0} = abV_o$$
$$V_{t=2} = abV_o u^{N-i} d^i \ (i = 0, 1, 2, \ldots, N)$$
$$V_{t=4} = abV_o u^{2N-i} d^i \ (i = 0, 1, 2, \ldots, 2 \cdot N)$$
$$V_{t=6} = abV_o u^{3N-i} d^i \ (i = 0, 1, 2, \ldots, 3 \cdot N)$$
$$V_{t=8} = abV_o u^{4N-i} d^i \ (i = 0, 1, 2, \ldots, 4 \cdot N)$$
$$V_{t=10} = abV_o u^{5N-i} d^i \ (i = 0, 1, 2, \ldots, 5 \cdot N)$$

The next step is to evaluate real options based on a binomial lattice, which is already done by the first step. The call option value in terminal nodes and decision points can be calculated in the same way as a put option, described in Section 14.3. The values can be obtained by pursuing maximization between the value obtained by exercising a call option and the zero value obtained by abandoning the option. Figure 14.4 describes the procedures for obtaining a call option premium from terminal nodes.

- Terminal node:

$$C_{t=10}^{i_5} = \max(abV_5 Z_5 - K_5, 0)$$

$$= \max(abV_0 u^{5N-m} d^m Z_5 - k_5, 0) \qquad (m = 0, 1, 2, \ldots, 5 \cdot N)$$

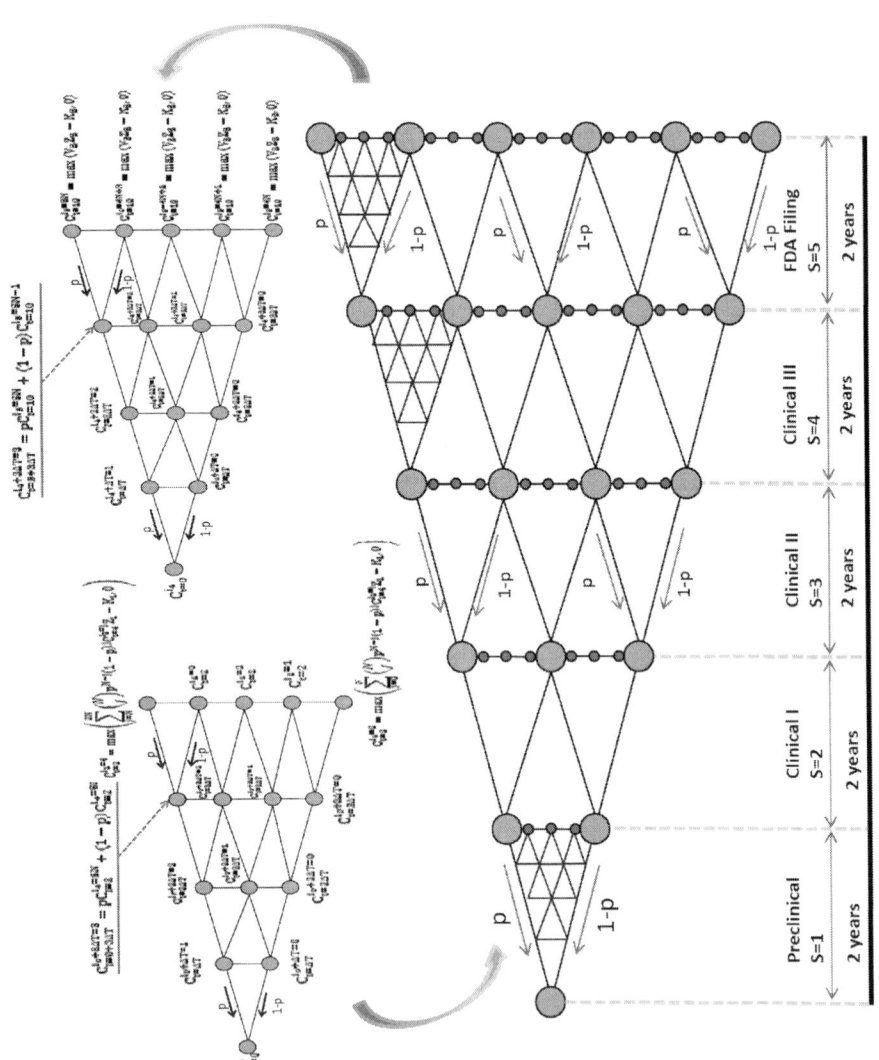

FIGURE 14.4 Valuation lattice.

- Intermediate decision point:

$$C_{t=8}^{i_4=m} = \max\left(\sum_{j=m}^{N+m}\binom{N}{j}p^{N-j}(1-p)^{j}C_{t=10}^{i_5=j}Z_4 - K_4, 0\right) \quad (m=0, 1, 2, \ldots, 4\cdot N)$$

$$C_{t=6}^{i_3=m} = \max\left(\sum_{j=m}^{N+m}\binom{N}{j}p^{N-j}(1-p)^{j}C_{t=8}^{i_4=j}Z_3 - K_3, 0\right) \quad (m=0, 1, 2, \ldots, 3\cdot N)$$

$$C_{t=4}^{i_2=m} = \max\left(\sum_{j=m}^{N+m}\binom{N}{j}p^{N-j}(1-p)^{j}C_{t=6}^{i_3=j}Z_2 - K_2, 0\right) \quad (m=0, 1, 2, \ldots, 2\cdot N)$$

$$C_{t=2}^{i_1=m} = \max\left(\sum_{j=m}^{N+m}\binom{N}{j}p^{N-j}(1-p)^{j}C_{t=4}^{i_2=j}Z_1 - K_1, 0\right) \quad (m=0, 1, 2, \ldots, N)$$

- Call option premium: $C_{t=0}^{i_0} = \sum_{j=0}^{N}\binom{N}{j}p^{N-j}(1-p)^{j}C_{t=2}^{i_1=j}$

Once the call option premium is calculated, pharmaceutical companies can decide how to react to a licensing problem. If a call option is positive, the company may be willing to propel this licensing agreement forward. Otherwise, the company will seek other alternatives because this licensing agreement will be financially disadvantageous.

14.5   OPTIMAL PARTNERSHIP TIMING STRATEGY

Based on a real options approach, the consensus of a contract can be viewed as applying a call option for the pharmaceutical company and can be viewed as applying a put option for a biotechnology company. As described in Figure 14.5, pharmaceutical companies pursue later partnerships when some risks are eliminated to lower contract payments and make smaller up-front payments. On the other hand, biotechnology companies seek earlier partnerships to obtain secured payment even in the eventuality of abandoning a project.

Since the direction for exercising an option that each company pursues in licensing is totally opposite, an anticipated overlapping zone that satisfies both parties is possible, as shown in Figure 14.5. This overlapping region is formed by the condition of a positive call option premium and a positive put option premium guaranteeing that a licensing agreement will not damage either company, even in the worst-case scenario of failure, if an alliance is achieved in this overlapping region. Thus, determining the optimal timing occurs when the sum of call option premiums and put option premiums reaches a maximum value.

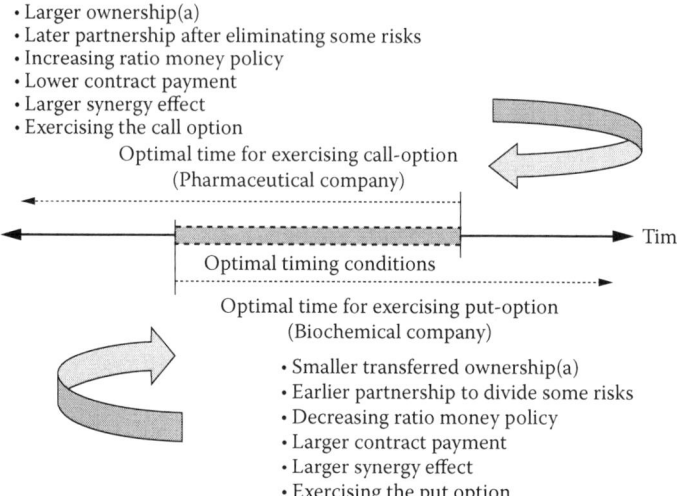

- Larger ownership(a)
- Later partnership after eliminating some risks
- Increasing ratio money policy
- Lower contract payment
- Larger synergy effect
- Exercising the call option

Optimal time for exercising call-option
(Pharmaceutical company)

Optimal timing conditions

Time

Optimal time for exercising put-option
(Biochemical company)

- Smaller transferred ownership(a)
- Earlier partnership to divide some risks
- Decreasing ratio money policy
- Larger contract payment
- Larger synergy effect
- Exercising the put option

FIGURE 14.5 Summary of the partnership contract.

14.6   CONCLUSION

The real options framework can be useful for project valuation, especially in new drug development. Unlike general R&D projects, new-drug R&D projects have to pass through five well-defined steps. R&D projects in medical fields incur substantial R&D expenses and take significant amounts of time to complete all the developmental stages. Failure halfway along could return nothing to the investor; thus, consideration for decisions at every point in time is essential.

The study introduces a project evaluation methodology based on a binomial decision tree. The partnerships' main assets arise from the value of the project. A partnership's contract is regarded as a call option for a pharmaceutical company because partnership means buying part of the ownership from the biotechnology company. In the instance of a biotechnology company, the contract means exercising a put option to sell ownership. Throughout this real options base, the search is for the optimal timing range that could satisfy both companies. Having determined that range, the next task is determining the optimal timing to reach maximum value throughout the partnership on the basis of the sum of the option premiums for the two firms.

If all assumed parameters are provided, this methodology will suggest optimal partnership timing as well as subsequent conditions in a successful partnership. These consecutive conditions embrace the ownership ratio, money policy, and amount of milestone payment at every stage that can be consequently determined when the partnership is achieved in optimal environment.

In some sense, attaining partnership conditions and timing is a part of the numerical example based on presuggested parameters. Many parameters practically are very sensitive to the fluctuating market and technical factors. It may be difficult to set these parameters as fixed numbers in a practical view. Thus, more realistic and better approaches are required to determine the effects of parameters. Sensitivity analysis is a plausible approach to recognize the power of parameters' influence by creating a set of scenarios within specific boundaries. Its comparative approach may be helpful for project managers involving a partnership alliance to determine how changes in each volatile variable will impact timing as well as the subsequent conditions when a partnership was made. Kim (2009) regards as important parameters the synergy effects ratio, technical success ratio, market volatility, remaining R&D expenses, and sensitivity analysis with which these parameters is studied.

Ultimately, the real options approach within this framework provides a blueprint for examining optimal timing strategies for partnerships.

REFERENCES

Arnold, K., Coia, A., Saywell, S., Smith, T., Minick, S., and Löffler, A. 2002. Value drivers in licensing deals. http://www.kellogg.northwestern.edu/academic/biotech/faculty/onbiotech.htm.

Brennan M. J. and Schwartz, E. S. 1985. Evaluating Natural Resource Investments. *The Journal of Business* 58(2): 135–157.

Clarke, E. and Rousseau, P. 2002. Strategic parameters for capital budgeting when abandonment value is stochastic. *Applied Financial Economic* 12: 123–130.

Cohen, J., Gangi, W., Lineen, J., and Manard, A. 2005. Strategic alternatives in the pharmaceutical industry. http://www.kellogg.northwestern.edu/academic/biotech/faculty/onbiotech.htm.

Cox, J., Ross, S., and Rubinstein, M. 1979. Option pricing: A simplified approach. *Journal of Financial Economics* 7:229–264.

Dixit, A. K. 1989. Entry and Exit Decisions under Uncertainty. *TheJournal of Political Economy* 97(3): 620–638.

Dixit, A. K., and Pindyck, R. S. 1994. Investments under uncertainty. Princeton, NJ: Princeton University Press.

Hagedoorn, J. 2002. Inter-firm R&D partnerships: An overview of major trends and patterns since 1960. *Research Policy* 31:477–492.

Herath, H. S. B., and Park, C. S. 1999. Economic analysis of R&D projects: An options approach. *The Engineering Economist* 44(1): 1–35.

Herath, H. S. B., and Park, C. S. 2002. Multi-stage capital investment opportunities as compound real options. *The Engineering Economist* 47(1): 1–27.

Kim, M. J. 2009. Partnership conditions of a new drug development based on the real option. Master's thesis,

McDonald, R. and Siegel, D. 1986. The Value of Waiting to Invest. *The Quarterly Journal of Economics* 101 (4): 707–728.

Mitchell, G. R. and Nampton, W. F. 1988. Managing R&D as a strategic option. *Research-Technology Management* May-June: 15–22.

Myers, S. C. 1984. Finance Theory and Financial Strategy. *Strategic Management* 14 (Jan-Feb): 126–137.

Nichols, N. 1994. Scientific management at Merck: An interview with CFO Judy Lewent. *Harvard Business Review* 72:89–99.

Park, C. S., and Herath, H. S. B. 2000. Exploiting uncertainty-investment opportunities as real options: A new way of thinking in engineering economics. *The Engineering Economist* 45(1): 1–36.

Rogers, M. J., Gupta, A., and Maranas, C. D. 2002. Real options based analysis of optimal pharmaceutical research and development portfolios. *Industrial and Engineering Chemistry Research* 41(25): 6607–6620.

Rogers, M. J., Maranas C. D., and Ding, M. 2005. Valuation and design of pharmaceutical companies R&D licensing deals. *AICHE Journal* 51(1): 198–209.

Smith, R. L. 2001. Strategic and tactical perspectives of drug delivery alliances with Big Pharma. *American Pharmaceutical Review* 4(2): 81–84.

15 Hands-On Applications
Real Option Super Lattice Solver Software

Johnathan Mun
Real Options Valuation, Inc.

CONTENTS

Once a real options framework has been *defined* for a problem or a project, the true goal is to *value* that option. The use of software-based models allows the analyst to apply a consistent, well-tested, and replicable set of models. It reduces computational errors and allows the user to focus more on the process and problem at hand rather than on building potentially complex and mathematically intractable models. This chapter provides a good starting point with an introduction to the Real Options Super Lattice Solver software. For more details on using the software, consult the user manual, whereas for more technical, theoretical, and practical details of real options analysis, consult *Real Options Analysis: Tools and Techniques,* 2nd ed. (Wiley Finance; Mun 2005). The materials covered in this chapter assume that the reader is sufficiently well versed in the basics of real options analytics.

15.1   ACQUIRING SLS SOFTWARE

To get started using the software, visit **www.realoptionsvaluation.com** and click on the **Downloads** link to view some free **Getting Started** videos on advanced risk analysis and real options modeling techniques using **Risk Simulator** and **Real Options SLS** software applications, plus to download the trial software versions and sample models (you can also watch many free **Getting Started** videos, get free models at this site, and try out other powerful software like Risk Simulator for running Monte Carlo simulations, stochastic forecasting, and portfolio optimizations as a prerequisite for running real options models). Please review the user manuals for step-by-step instructions, sample models, and solved case studies and applications.

15.2   INTRODUCTION TO REAL OPTIONS
SUPER LATTICE SOLVER SOFTWARE

The Real Options Super Lattice (SLS) software comprises several modules, including the Single Super Lattice Solver (SLS), Multiple Super Lattice Solver (MSLS), Multinomial Lattice Solver (MNLS), SLS Excel Solution, SLS Functions, and Exotic Options Valuator. These modules are highly powerful and customizable binomial and multinomial lattice solvers and can be used to solve many types of options (including the three main families of options: real options, which deals with physical and intangible assets; financial options, which deals with financial assets and the investments of such assets; and employee stock options, which deals with financial assets provided to employees within a corporation). This text illustrates some sample real options, financial options, and employee stock options applications that users will encounter most frequently. The following are the modules in the Real Options Super Lattice software:

- The **SLS** is used primarily for solving options with a *single underlying asset* using binomial lattices. Even highly complex options with a single underlying asset can be solved using the SLS. The types of options solved include American, Bermudan, and European options to abandon, choose, contract, defer, execute, expand, and wait, as well as any customized combinations of these options with changing inputs.
- The **MSLS** is used for solving options with *multiple underlying assets* and sequential compound options with *multiple phases* using binomial lattices. Highly complex options with multiple underlying assets and phases can be solved using the MSLS. The types of options solved include multiple-phased stage-gate sequential compound options, simultaneous compound options, switching options, multiple-asset chooser options, and customized combinations of phased options with all the option types solved using the SLS module described above.
- The **MNLS** uses *multinomial lattices* (trinomial, quadranomial, and pentanomial) to solve specific options that cannot be solved using binomial lattices. The options solved include mean-reverting, jump diffusion, and rainbow options.

- The **SLS Excel Solution** implements the SLS and MSLS computations within the Excel environment, allowing users to access the SLS and MSLS functions directly in Excel. This feature facilitates model building, formula and value linking and embedding, as well as running simulations, and provides the user with sample templates to create such models.
- The **SLS Functions** are additional real options and financial options models accessible directly through Excel. This module facilitates model building, linking and embedding, and running simulations.
- The **Exotic Options Valuator** has over 300 exotic, financial, employee, and real options models as well as options-related models (e.g., bond options, hedge ratios, volatility, and others) available for your use in a simple tool.

The SLS software is created by me and accompanies the materials presented at different training courses on real options, simulation, employee stock options valuation, and Certified in Risk Management (CRM) programs taught by me. While the software and its models are based on my books, the training courses cover the real options subject matter in more depth, including the solution of sample business cases and the framing of real options of actual cases. It is highly suggested that the reader familiarizes him- or herself with the fundamental concepts of real options in Chapters 6 and 7 of *Real Options Analysis,* 2nd ed. (Mun 2005), prior to attempting an in-depth real options analysis using the software. Note that the first edition of *Real Options Analysis: Tools and Techniques* (Mun 2002) shows the Real Options Analysis Toolkit software, an older precursor to the Super Lattice Solver, also created by me. The Super Lattice Solver supersedes the Real Options Analysis Toolkit by providing the following enhancements, and is introduced in the second edition of *Real Options Analysis:*

- All inconsistencies, computation errors, and bugs are fixed and verified.
- Allowance of changing input parameters over time (customized options).
- Allowance of changing volatilities over time.
- Incorporation of American, Asian, Bermudan, European (vesting and blackout periods), and customized options.
- Flexible modeling capabilities in creating or engineering your own customized options.
- General enhancements to accuracy, precision, and analytical prowess.
- Addition of over 300 exotic options models and options-related models.

As the creator of both the Super Lattice Solver and Real Options Analysis Toolkit software, I suggest that the reader focuses on using the Super Lattice Solver as it provides many powerful enhancements and analytical flexibility over its predecessor, the older, less powerful, and less flexible Real Options Analysis Toolkit software.

15.3   SINGLE ASSET SUPER LATTICE SOLVER (SLS)

Figure 15.1 illustrates the SLS module. After installing the software, the user can access the SLS by clicking on **Start | Programs | Real Options Valuation | Real Options SLS | Real Options SLS**. The SLS has several sections: *Option Type, Basic Inputs, Custom Equations, Custom Variables, Benchmark, Result, and Create Audit Worksheet.*

15.3.1   SLS EXAMPLES

To help you get started, several simple examples are in order. A simple European call option is computed in this example using SLS. To follow along, in the software, start this example file by selecting File | Sample Files | Plain Vanilla Call Option I. This example file will be loaded into the SLS software, as seen in Figure 15.2. The starting PV Underlying Asset or starting stock price is $100, and the Implementation Cost or strike price is $100 with a five-year maturity. The annualized risk-free rate of return is 5 percent, and the historical, comparable, or future expected annualized volatility is 10 percent. Click on RUN (or Alt-R), and a 100-step binomial lattice is

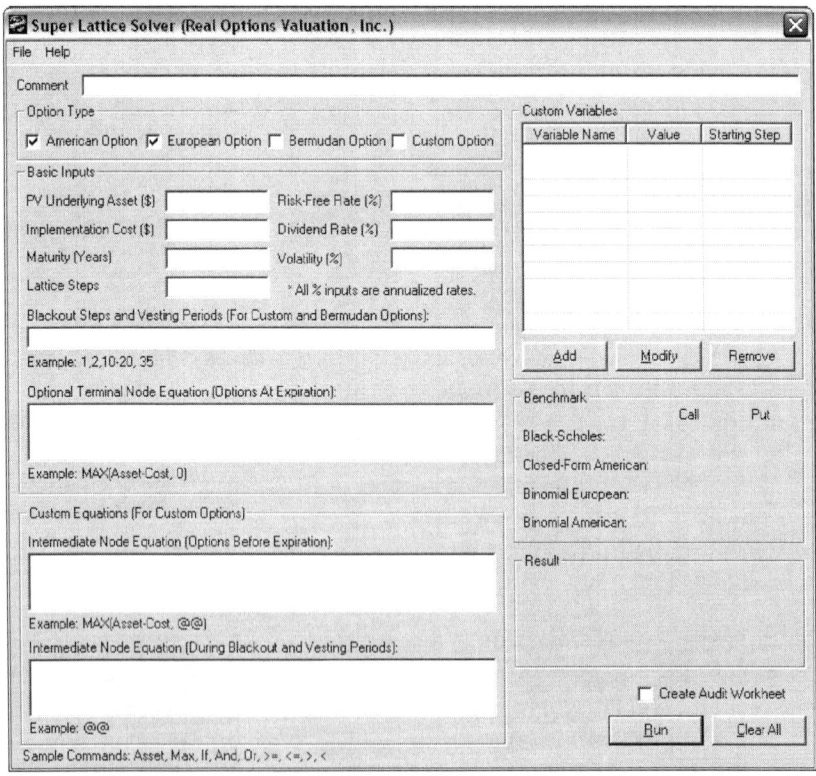

FIGURE 15.1 Single Super Lattice Solver (SLS).

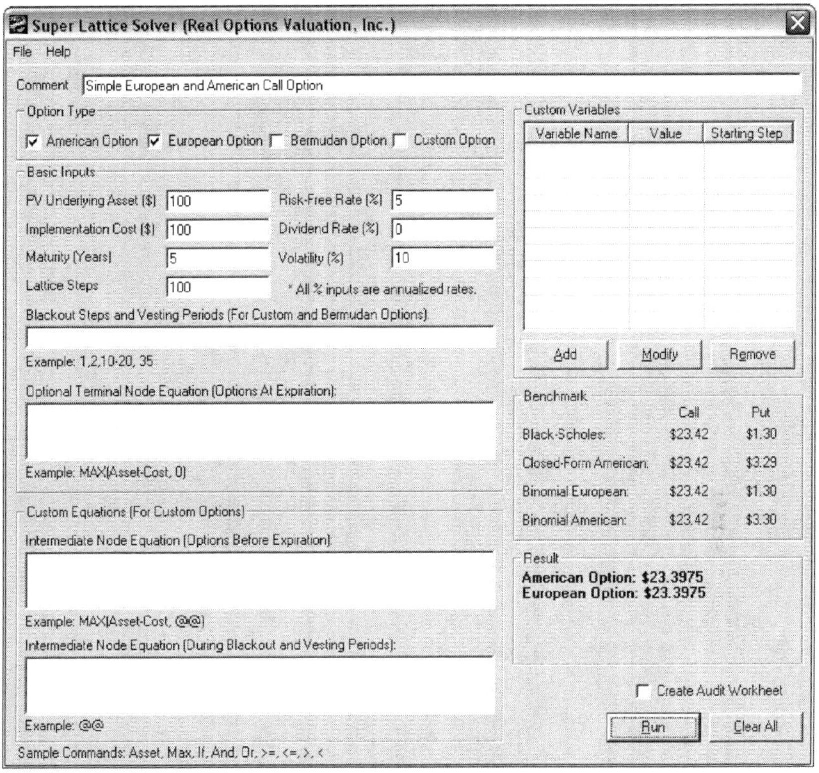

FIGURE 15.2 SLS results of a simple European and American call option.

computed and the results indicate a value of $23.3975 for both the European and American call options. Benchmark values using the Black-Scholes and closed-form American approximation models as well as standard plain vanilla binomial American and binomial European call and put options with 1,000-step binomial lattices are also computed. Notice that only the American and European options are selected and that the computed results are for these simple plain vanilla American and European call options.

The benchmark results use both closed-form models (Black-Scholes and closed-form approximation models) and 1,000-step binomial lattices on plain vanilla options. You can change the steps to 1,000 in the basic inputs section to verify that the answers computed are equivalent to the benchmarks, as seen in Figure 15.3. Notice that, of course, the values computed for the American and European options are identical to each other and identical to the benchmark values of $23.4187, as it is never optimal to exercise a standard plain vanilla call option early if there are no dividends. Be aware that the higher the lattice steps, the longer it takes to compute the results. It is advisable to start with lower lattice steps to make sure the analysis is robust, and then progressively increase lattice steps to check for results convergence.

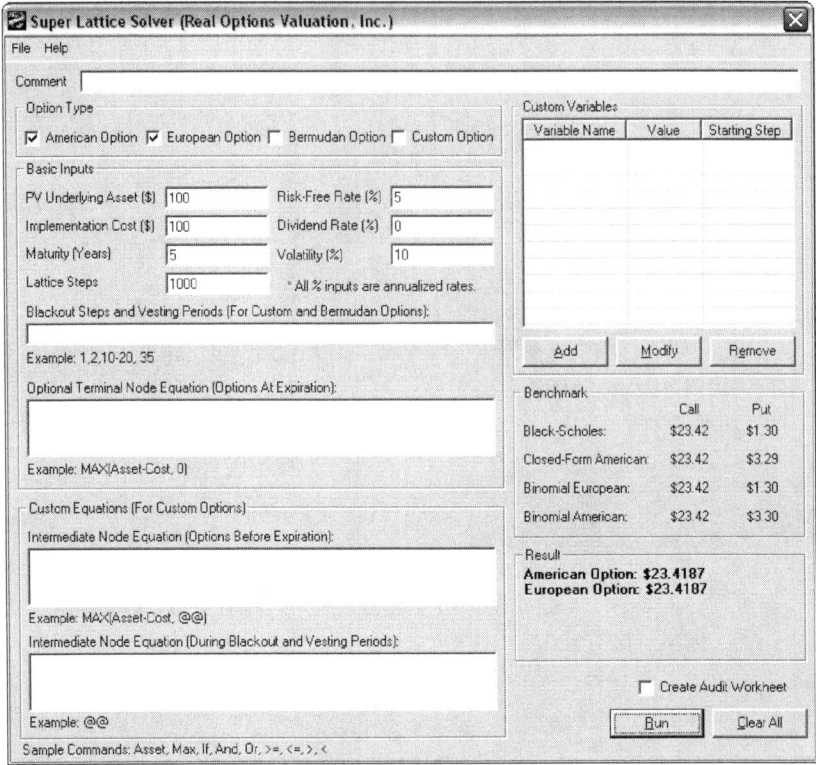

FIGURE 15.3 SLS comparing results with benchmarks.

Alternatively, you can enter Terminal and Intermediate Node Equations for a call option to obtain the same results. Notice that using 100 steps and creating your own Terminal Node Equation of **Max(Asset-Cost,0)** and Intermediate Node Equation of **Max(Asset-Cost,OptionOpen)** will yield the same answer. When entering your own equations, make sure that Custom Option is first checked.

Figure 15.4 illustrates how the analysis is done. The example file used in this illustration is "Plain Vanilla Call Option III." Notice that the value $23.3975 in Figure 15.4 agrees with the value in Figure 15.2. The Terminal Node Equation is the computation that occurs at maturity, while the Intermediate Node Equation is the computation that occurs at all periods prior to maturity, and is computed using backward induction. The term "OptionOpen" represents "keeping the option open," and is often used in the Intermediate Node Equation when analytically representing the fact that the option is not executed but kept open for possible future execution. Therefore, in Figure 15.4, the Intermediate Node Equation **Max(Asset-Cost, OptionOpen)** represents the profit maximization decision of either executing the option or leaving it open for possible future execution. In contrast, the Terminal Node Equation of **Max(Asset-Cost,0)** represents the profit maximization decision at maturity of either executing the option if it is in the money, or allowing it to expire worthless if it is at the money or out of the money.

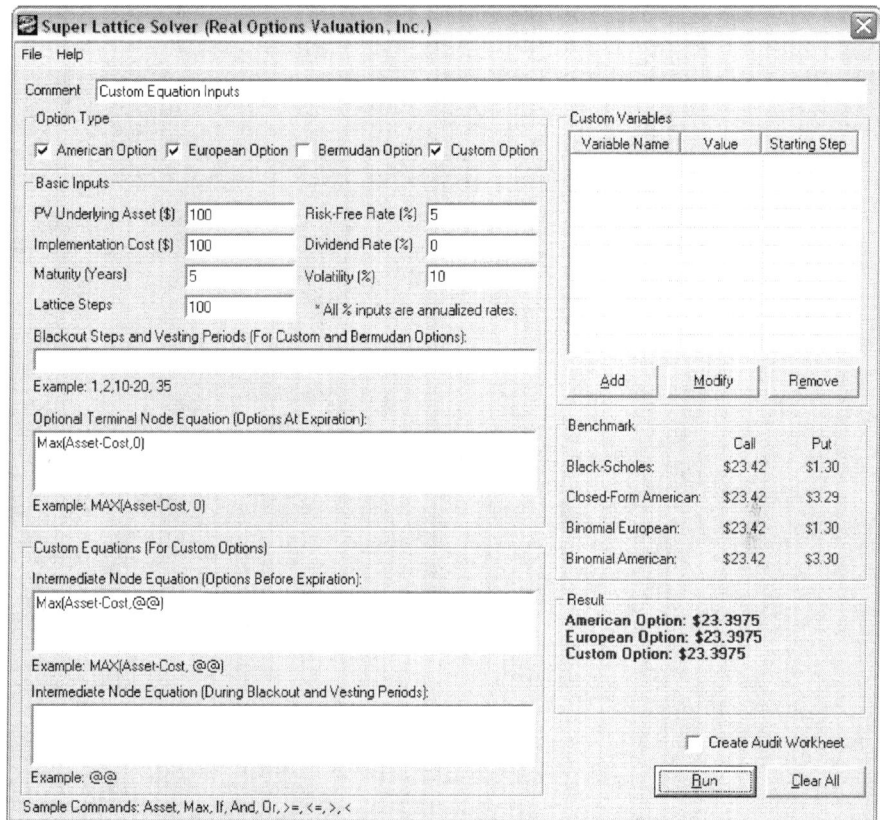

FIGURE 15.4 Custom equation inputs.

When entering your own equations, make sure that Custom Option is first checked.

In addition, you can create an audit worksheet in Excel to view a sample ten-step binomial lattice by checking the box Create Audit Worksheet. For instance, loading the example file "Plain Vanilla Call Option I" and selecting the box create a worksheet, as seen in Figure 15.5. Several items on this audit worksheet are noteworthy:

- The audit worksheet generated will show the first ten steps of the lattice, regardless of how many you enter. That is, if you enter 1,000 steps, the first ten steps will be generated. If a complete lattice is required, simply enter ten steps in the SLS and the full ten-step lattice will be generated instead. The Intermediate Computations and Results are for the Super Lattice, based on the number of lattice steps entered, and not based on the ten-step lattice generated. To obtain the Intermediate Computations for ten-step lattices, simply rerun the analysis inputting 10 as the lattice steps. This way, the audit worksheet generated will be for a ten-step lattice, and the results from SLS will now be comparable (Figure 15.6).

Option Valuation Audit Sheet

Assumptions		Intermediate Computations	
PV Asset Value ($)	$100.00	Stepping Time (dt)	0.0500
Implementation Cost ($)	$100.00	Up Step Size (up)	1.0226
Maturity (Years)	5.00	Down Step Size (down)	0.9779
Risk-free Rate (%)	5.00%	Risk-neutral Probability	0.5504
Dividends (%)	0.00%		
Volatility (%)	10.00%	**Results**	
Lattice Steps	100	Lattice Result	23.40
Option Type	European		

Terminal Equation	Max(Asset-Cost,0)
Intermediate Equation	@@
Intermediate Equation (Blackouts)	@@

Underlying Asset Lattice

								125.06
							122.29	
						119.59		119.59
					116.94		116.94	
				114.36		114.36		114.36
			111.83		111.83		111.83	
		109.36		109.36		109.36		109.36
	106.94		106.94		106.94		106.94	
104.57		104.57		104.57		104.57		104.57
102.26		102.26		102.26		102.26		
100.00	100.00		100.00		100.00		100.00	100.00
97.79		97.79		97.79		97.79		
	95.63		95.63		95.63		95.63	95.63
	93.51		93.51		93.51		93.51	
		91.44		91.44		91.44		91.44
		89.42		89.42		89.42		
			87.44		87.44		87.44	87.44
				85.51		85.51		
					83.62		83.62	83.62
						81.77		
								79.96

Option Valuation Lattice

								45.33
							42.81	
						40.35		39.96
					37.97		37.58	
				35.66		35.27		34.87
			33.43		33.04		32.64	
		31.27		30.88		30.49		30.09
	29.18		28.80		28.41		28.02	
27.18		26.79		26.41		26.02		25.64
25.25		24.87		24.49		24.11		23.73
23.40	23.03		22.65		22.28		21.90	21.52
21.26		20.90		20.53		20.16		19.79
	19.22		18.86		18.50		18.14	17.77
	17.28		16.93		16.58		16.22	
		15.45		15.10		14.76		14.41
		13.71		13.38		13.05		
			12.09		11.77		11.45	11.45
				10.58		10.27		
					9.19		8.89	8.89
						7.91		
								6.74

FIGURE 15.5 SLS generated audit worksheet.

- The worksheet only provides values as it is assumed that the user was the one who entered in the terminal and intermediate node equations; hence, there is really no need to recreate these equations in Excel. The user can always reload the SLS file and view the equations or print out the form if required (by clicking on **File | Print**).

The software also allows you to save or open analysis files. That is, all the inputs in the software will be saved and can be retrieved for future use. The results will not be saved

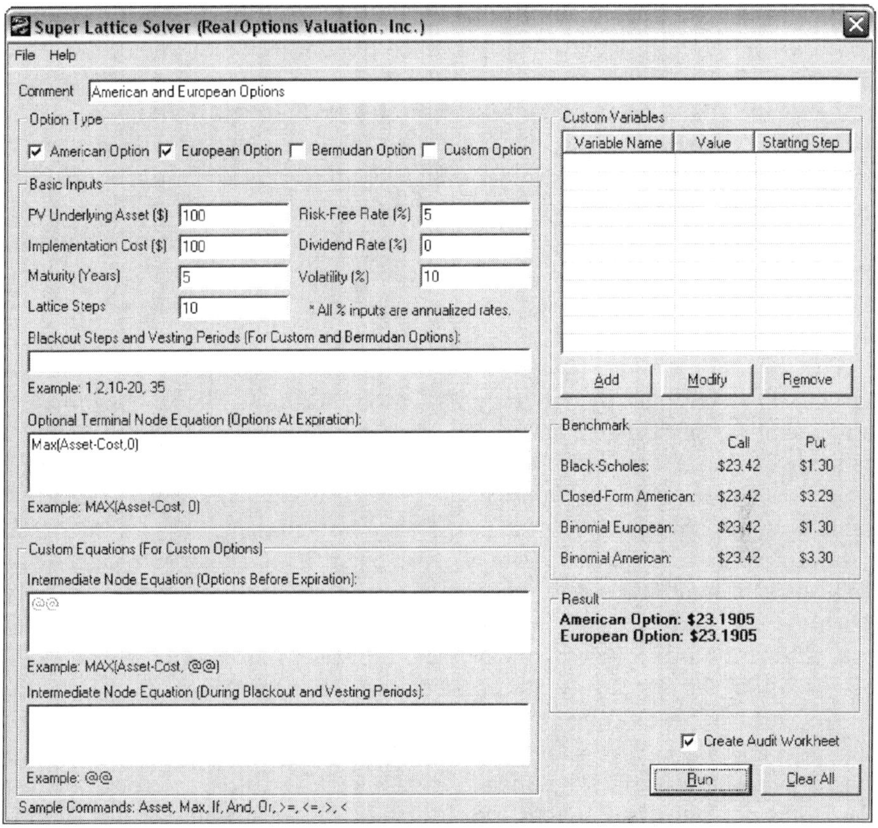

FIGURE 15.6 SLS results with a ten-step lattice.

because you may accidentally delete or change an input and the results will no longer be valid. In addition, rerunning the super lattice computations will take only a few seconds, and it is advisable for you to always rerun the model when opening an old analysis file.

You may also enter in Blackout Steps. These are the steps on the super lattice that will have different behaviors than the terminal or intermediate steps. For instance, you can enter "1000" as the lattice steps, and enter "0-400" as the blackout steps, and some Blackout Equation (e.g., OptionOpen). This means that for the first 400 steps, the option holder can only keep the option open. Other examples include entering "1, 3, 5, 10" if these are the lattice steps where blackout periods occur. You will have to calculate the relevant steps within the lattice where the blackout exists. For instance, if the blackout exists in years 1 and 3 on a ten-year, ten-step lattice, then steps 1, 3 will be the blackout dates. This blackout step feature comes in handy when analyzing options with holding periods, vesting periods, or periods where the option cannot be executed. Employee stock options have blackout and vesting periods, and certain contractual real options have periods during which the option cannot be executed (e.g., cooling-off periods, or proof-of-concept periods).

If equations are entered into the Terminal Node Equation box and American, European, or Bermudan Options are chosen, the Terminal Node Equation you entered will be the one used in the super lattice for the terminal nodes. However, for the intermediate nodes, the American option assumes the same Terminal Node Equation plus the ability to keep the option open; the European option assumes that the option can only be kept open and not executed; while the Bermudan option assumes that during the blackout lattice steps, the option will be kept open and cannot be executed. If you also enter the Intermediate Node Equation, the Custom Option should first be chosen (otherwise you cannot use the Intermediate Node Equation box). The Custom Option result uses all the equations you have entered in Terminal, Intermediate, and Intermediate during Blackout Sections.

The Custom Variables list is where you can add, modify, or delete custom variables, the variables that are required beyond the basic inputs. For instance, when running an abandonment option, you need the salvage value. You can add this in the Custom Variables list, and provide it a name (a variable name must be a single word), the appropriate value, and the starting step when this value becomes effective. For example, if you have multiple salvage values (i.e., if salvage values change over time), you can enter the same variable name (e.g., "salvage") several times, but each time, its value changes and you can specify when the appropriate salvage value becomes effective. For instance, in a ten-year, 100-step super lattice problem where there are two salvage values—$100 occurs within the first five years and increases to $150 at the beginning of Year 6—you can enter two salvage variables with the same name, $100 with a starting step of 0, and $150 with a starting step of 51. Be careful here as Year 6 starts at step 51 and not 61. That is, for a ten-year option with a 100-step lattice, we have the following: Steps 1–10 = Year 1; Steps 11–20 = Year 2; Steps 21–30 = Year 3; Steps 31–40 = Year 4; Steps 41–50 = Year 5; Steps 51–60 = Year 6; Steps 61–70 = Year 7; Steps 71–80 = Year 8; Steps 81–90 = Year 9; and Steps 91–100 = Year 10. Finally, incorporating "0" as a blackout step indicates that the option cannot be executed immediately.

15.4   MULTIPLE SUPER LATTICE SOLVER (MSLS)

The MSLS is an extension of the SLS in that the MSLS can be used to solve options with multiple underlying assets and multiple phases. The MSLS allows the user to enter multiple underlying assets as well as multiple valuation lattices (Figure 15.7). These valuation lattices can call to user-defined custom variables. Some examples of the types of options that the MSLS can be used to solve include the following:

- Sequential compound options (two-, three-, and multiple-phased sequential options)
- Simultaneous compound options (multiple assets with multiple simultaneous options)
- Chooser and switching options (choosing among several options and underlying assets)

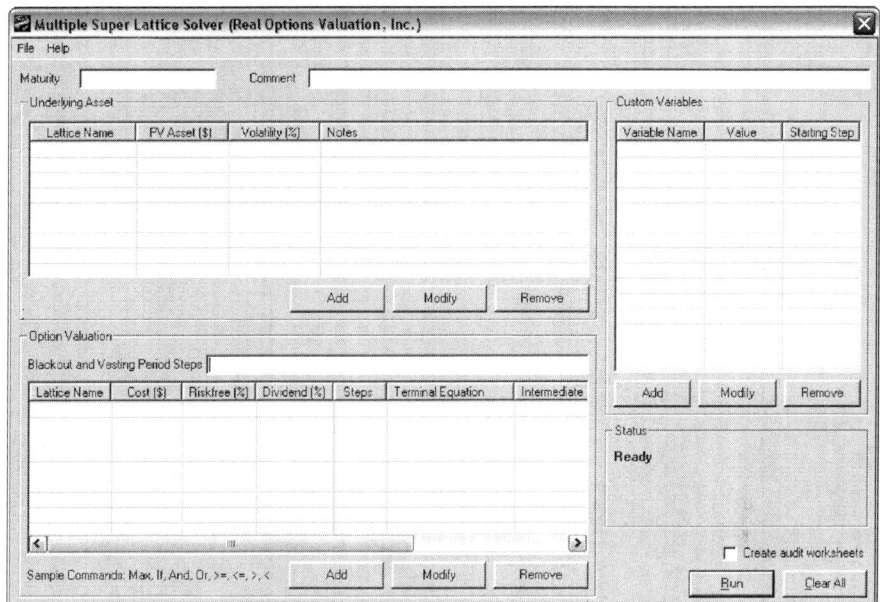

FIGURE 15.7 Multiple Super Lattice Solver.

- Floating options (choosing between calls and puts)
- Multiple asset options (3D binomial option models)

The MSLS software has several areas including a Maturity and Comment area. The Maturity value is a global value for the entire option, regardless of how many underlying or valuation lattices exist. The comment field is for your personal notes describing the model you are building. MSLS includes a Blackout and Vesting Period Steps section and a Custom Variables list similar to the SLS. The MSLS also allows you to create audit worksheets.

To illustrate the power of the MSLS, a simple illustration is in order. In the MSLS module, click on **File | Sample Files | MSLS** – Two Phased Sequential **Compound Option** to open this example. Figure 15.8 shows the MSLS example loaded. In this simple example, a single underlying asset is created with two valuation phases.

The strategy tree for this option is seen in Figure 15.9. The project is executed in two phases—the first phase within the first year costs $5 million, while the second phase occurs within two years but only after the first phase is executed, and costs $80 million, both in present-value (PV) dollars. The PV asset of the project is $100 million (the net present value [NPV] is therefore $15 million), and faces 30 percent volatility in its cash flows. The computed strategic value using the MSLS is $27.67 million, indicating that there is $12.67 million in option value. That is, spreading out and staging the investment into two phases have significant value (an expected value of $12.67 million, to be exact).

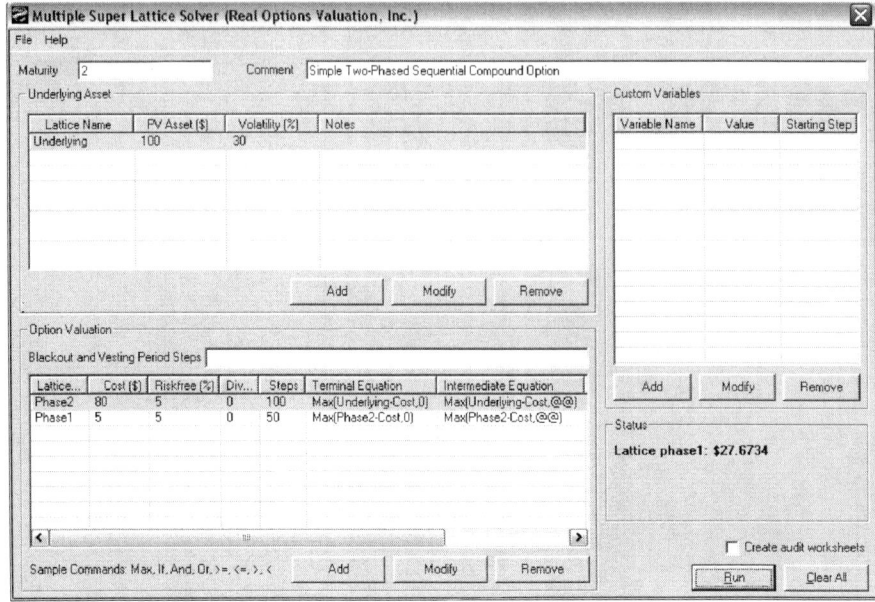

FIGURE 15.8 MSLS solution to a simple two-phased sequential compound option.

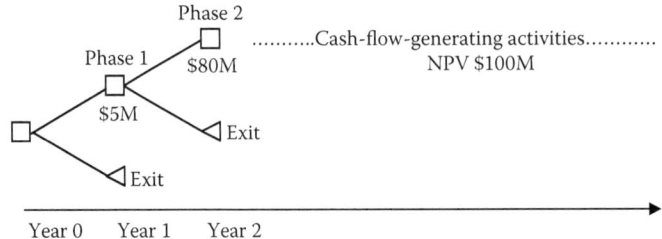

FIGURE 15.9 Strategy tree for two-phased sequential compound option.

15.5   MULTINOMIAL LATTICE SOLVER (MNLS)

The Multinomial Lattice Solver (MNLS) is another module of the Real Options Valuation's Super Lattice Solver software. The MNLS applies multinomial lattices—where multiple branches stem from each node—such as trinomials (three branches), quadranomials (four branches), and pentanomials (five branches). Figure 15.10 illustrates the MNLS module. The module has a Basic Inputs section, where all of the common inputs for the multinomials are listed. Then, there are four sections with four different multinomial applications complete with the additional required inputs and results for both American and European call and put options.

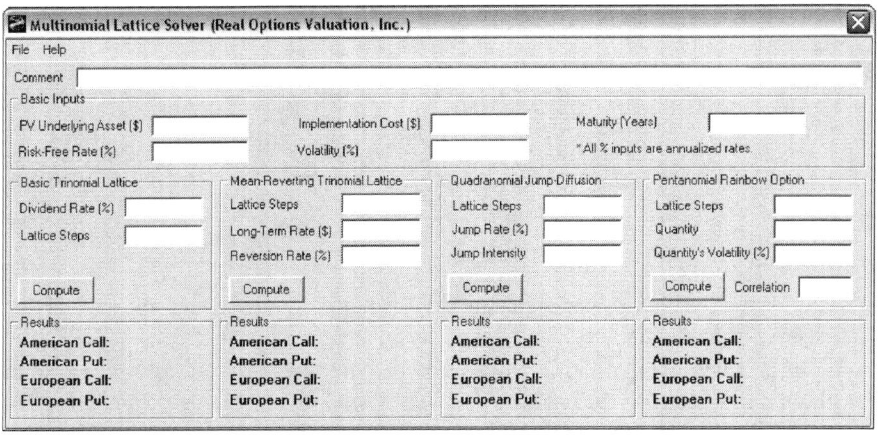

FIGURE 15.10 Multinomial Lattice Solver.

Figure 15.11 shows an example call and put option computation using trinomial lattices. To follow along, open the example file "MNLS—Simple Calls and Puts Using Trinomial Lattices." Note that the results shown in Figure 15.11 using a fifty-step lattice are equivalent to the results shown in Figure 15.2 using a 100-step binomial lattice. In fact, a trinomial lattice or any other multinomial lattice provides identical answers to the binomial lattice at the limit, but convergence is achieved faster at lower steps.

To illustrate, Table 15.1 shows how the trinomial lattice of a certain set of input assumptions yields the correct option value with fewer steps than it takes for a binomial lattice. Because both yield identical results at the limit but trinomials are much more difficult to calculate and take a longer computation time, the binomial lattice

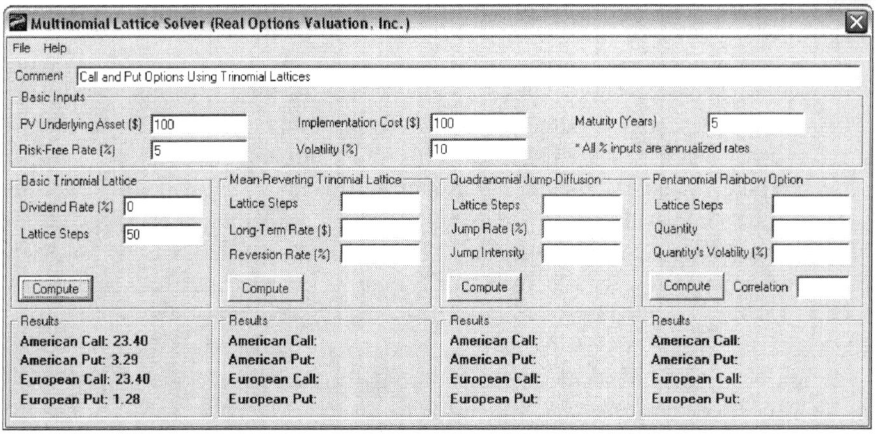

FIGURE 15.11 A simple call and put using trinomial lattices.

TABLE 15.1
Binomial versus Trinomial Lattices

Steps	5	10	100	1,000	5,000
Binomial lattice	$30.73	$29.22	$29.72	$29.77	$29.78
Trinomial lattice	$29.22	$29.50	$29.75	$29.78	$29.78

is usually used instead. However, a trinomial is required only under one special circumstance: when the underlying asset follows a mean-reverting process.

With the same logic, quadranomials and pentanomials yield identical results as the binomial lattice with the exception that these multinomial lattices can be used to solve the following different special limiting conditions:

- *Trinomials*: Results are identical to binomials and are most appropriate when used to solve mean-reverting underlying assets.
- *Quadranomials*: Results are identical to binomials and are most appropriate when used to solve options whose underlying assets follow jump diffusion processes.
- *Pentanomials*: Results are identical to binomials and are most appropriate when used to solve two underlying assets that are combined, called rainbow options (e.g., price and quantity are multiplied to obtain total revenues, but price and quantity each follows a different underlying lattice with its own volatility but both underlying parameters could be correlated to one another).

15.5.1 SLS Excel Solution (SLS, MSLS, and Changing Volatility Models in Excel)

The SLS software also allows you to create your own models in Excel using customized functions. This functionality is important because certain models may require linking from other spreadsheets or databases, run certain Excel macros and functions, or need to be simulated, or inputs may change over the course of modeling your options. This Excel compatibility allows you the flexibility to innovate within the Excel spreadsheet environment. Specifically, the sample worksheet included in the software solves the SLS, MSLS, and Changing Volatility model.

To illustrate, Figure 15.12 shows a Customized Abandonment Option solved using SLS. The same problem can be solved using the *SLS Excel Solution* by clicking on **Start | Programs | Real Options Valuation | Real Options SLS | SLS Excel Solution**. The sample solution is seen in Figure 15.13. Notice the same results using the SLS versus the SLS Excel Solution file. You can use the template provided by simply clicking on File I Save As in Excel and use the new file for your own modeling needs.

Similarly, the MSLS can also be solved using the SLS Excel Solver. Figure 15.14 shows a complex multiple-phased sequential compound option solved using the SLS Excel Solver. The results shown here are identical to the results generated from the MSLS module (example file: "MSLS—Multiple Phased Complex Sequential

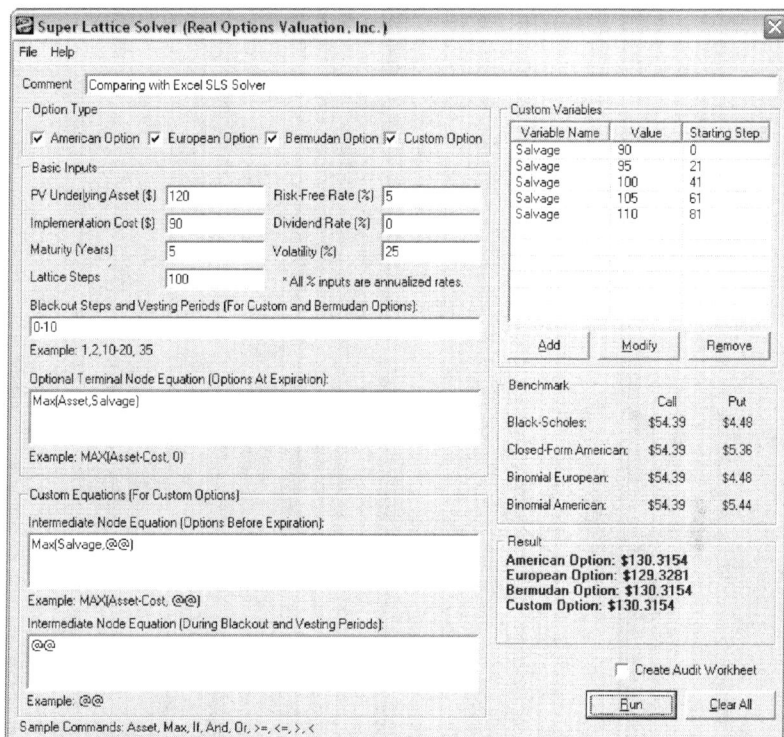

FIGURE 15.12 Customized abandonment option using SLS.

Compound Option"). One small note of caution here is that if you add or reduce the number of option valuation lattices, make sure you change the function's link for the MSLS Result to incorporate the right number of rows; otherwise, the analysis will not compute properly. For example, the default shows three option valuation lattices and by selecting the MSLS Results cell in the spreadsheet and clicking on Insert | Function, you

SUPER LATTICE SOLVER (SINGLE ASSET)

Option Type	0
PV Underlying Asset	$120.00
Annualized Volatility	25.00%
Maturity (Years)	5.00
Implementation Cost	$0.00
Risk-Free Rate	5.00%
Dividend Yield	0.00%
Lattice Steps	100
Terminal Equation	MAX(Asset, Salvage)
Intermediate Equation	MAX(Salvage, @@)
Intermediate Equation During Blackout	@@
Blackout Steps	0-10

Custom Variables List		
Variable Name	Value	Starting Steps
Salvage	90.00	0
Salvage	95.00	21
Salvage	100.00	41
Salvage	105.00	61
Salvage	110.00	81

Super Lattice Solver Result	$130.3154

Note: This is the Excel version of the Super Lattice Solver, useful when running simulations or when linking to and from other spreadsheets. Use this sample spreadsheet for your models. You can simply click on File, Save As to save as a different file and start using the model. For the option type, set 0 = American, 1 = European, 2 = Bermudan, 3 = Custom
The function used is: SLSSingle

FIGURE 15.13 Customized abandonment option using SLS Excel Solution.

MULTIPLE SUPER LATTICE SOLVER (MULTIPLE ASSET & MULTIPLE PHASES)

| Maturity (Years) | 5.00 |
| Blackout Steps | 0-20 |

MSLS Result $154.6802

Underlying Asset Lattices

Lattice Name	PV Asset	Volatility
Underlying	100.00	25.00

Custom Variables

Name	Value	Starting Steps
Salvage	100.00	31
Salvage	90.00	11
Salvage	80.00	0
Contract	0.90	0
Expansion	1.50	0
Savings	20.00	0

Option Valuation Lattices

Lattice Name	Cost	Riskfree	Dividend	Steps	Terminal Equation	Intermediate Equation	Intermediate Equation for Blackout
Phase3	50.00	5.00	0.00	50	Max(Underlying*Expansion-Cost,Underlying,Salvage)	Max(Underlying*Expansion-Cost,Salvage,@@)	@@
Phase2	0.00	5.00	0.00	30	Max(Phase3_Phase3*Contract+Savings,Salvage,0)	Max(Phase3*Contract+Savings,Salvage,@@)	@@
Phase1	0.00	5.00	0.00	10	Max(Phase2,Salvage,0)	Max(Salvage,@@)	@@

Note: This is the Excel version of the Multiple Super Lattice Solver, useful when running simulations or when linking to and from other spreadsheets.
Use this sample spreadsheet for your models. You can simply click on File, Save As to save as a different file and start using the model
The function used is: SLSMultiple
One small note of caution here is that if you add or reduce the number of option valuation lattices make sure you change the function's link for the MSLS Result
to incorporate the right number of rows otherwise the analysis will not compute properly. For example, the default shows 3 option valuation lattices and by selecting the MSLS Results
cell F5 and clicking on Insert I Function, you will see that the function links to cells A24:H26 for these three rows for the OVLattices input in the function. If you add another
option valuation lattice, change the link to A24:H27, and so forth.

FIGURE 15.14 Sequential compound (multi-phased) option using SLS Excel solution).

will see that the function links to cells A24:H26 for these three rows for the *OVLattices* input in the function. If you add another option valuation lattice, change the link to A24:H27, and so forth. You can also leave the list of custom variables as is. The results will not be affected if these variables are not used in the custom equations.

Finally, Figure 15.15 shows a Changing Volatility and Changing Risk-Free Rate Option. In this model, the volatility and risk-free yields are allowed to change over time, and a nonrecombining lattice is required to solve the option. In most cases, it is recommended that you create option models without the changing volatility term structure because getting a single volatility is difficult enough, let alone a series of changing volatilities over time. If different volatilities that are uncertain need to be modeled, run a Monte Carlo simulation using the Risk Simulator software on volatilities instead. This model should be used only when the volatilities are modeled robustly and the volatilities are rather certain and change over time. The same advice applies to a changing risk-free rate term structure.

15.5.2 SLS FUNCTIONS

The software also provides a series of SLS functions that are directly accessible in Excel. To illustrate its use, start the SLS Functions by clicking on **Start | Programs | Real Options Valuation | Real Options SLS | SLS Functions**, and Excel will start. When in Excel, you can click on the Function Wizard icon or simply select an empty cell and click on **Insert | Function**. While in Excel's Equation Wizard, select either the All category or **Real Options Valuation**, the name of the company that developed the software. Here you will see a list of SLS functions (with SLS prefixes) that are ready for use in Excel. Figure 15.16 shows the Excel Equation Wizard.

Suppose you select the first function, SLSBinomialAmericanCall, and hit OK. Figure 15.17 shows how the function can be linked to an existing Excel model. The values in cells B1 to B7 can be linked from other models or spreadsheets, can be created using Excel's Visual Basic for Applications (VBA) macros, or can be dynamic and changing as in when running a simulation. Another quick note of caution here is that certain SLS functions require many input variables, and Excel's Equation Wizard can show only five variables at a time. Therefore, remember to scroll down the list of variables by clicking on the vertical scroll bar to access the rest of the variables.

15.6 LATTICE MAKER

Next, the software also comes with an advanced binomial Lattice Maker module. This Lattice Maker is capable of generating binomial lattices and decision lattices with visible formulas in an Excel spreadsheet. Figure 15.18 illustrates an example option generated using this module. The illustration shows the module inputs (you can obtain this module by clicking on Lattice Maker in the main SLS menu) and the resulting output lattice. Notice that the visible equations are linked to the existing spreadsheet, which means this module will come in handy when running Monte Carlo simulations or when used to link to and from other spreadsheet models. The results can also be used as a presentation and learning tool to peep inside the analytical black box of binomial lattices. Last but not least, a decision lattice with specific

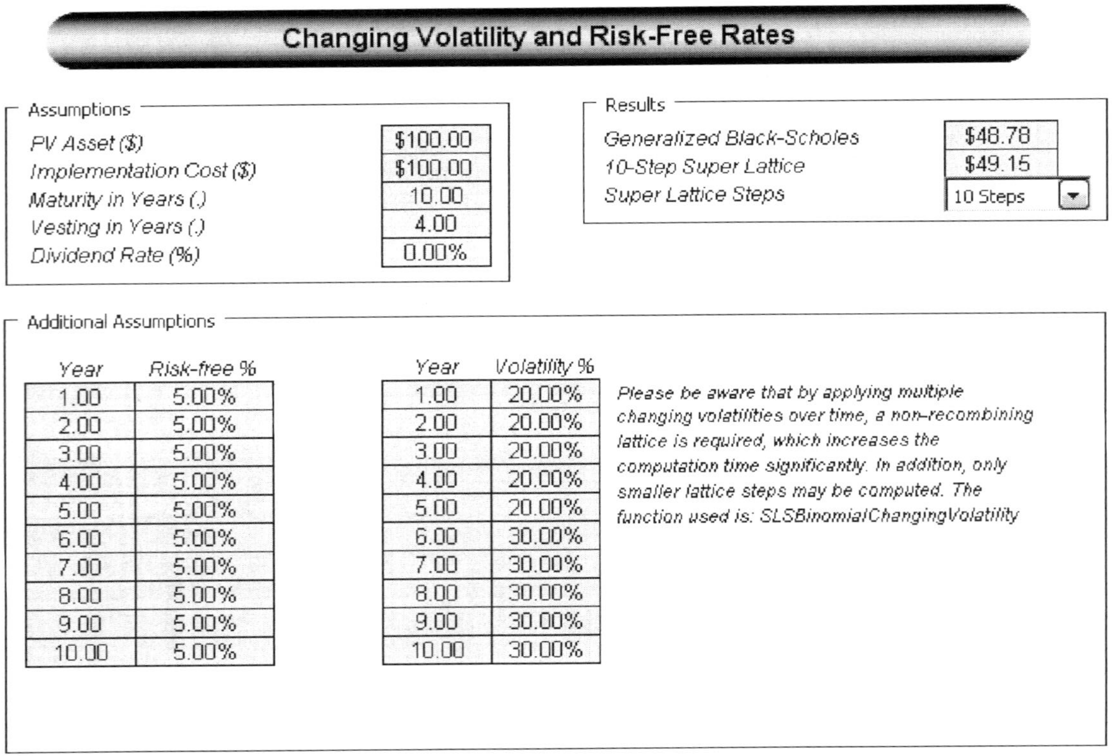

Changing Volatility and Risk-Free Rates

Results

Generalized Black-Scholes	$48.78
10-Step Super Lattice	$49.15
Super Lattice Steps	10 Steps ▶

Assumptions

PV Asset ($)	$100.00
Implementation Cost ($)	$100.00
Maturity in Years (.)	10.00
Vesting in Years (.)	4.00
Dividend Rate (%)	0.00%

Please be aware that by applying multiple changing volatilities over time, a non-recombining lattice is required, which increases the computation time significantly. In addition, only smaller lattice steps may be computed. The function used is: SLSBinomialChangingVolatility

Additional Assumptions

Year	Risk-free %
1.00	5.00%
2.00	5.00%
3.00	5.00%
4.00	5.00%
5.00	5.00%
6.00	5.00%
7.00	5.00%
8.00	5.00%
9.00	5.00%
10.00	5.00%

Year	Volatility %
1.00	20.00%
2.00	20.00%
3.00	20.00%
4.00	20.00%
5.00	20.00%
6.00	30.00%
7.00	30.00%
8.00	30.00%
9.00	30.00%
10.00	30.00%

FIGURE 15.15 Changing volatility and risk-free rate option.

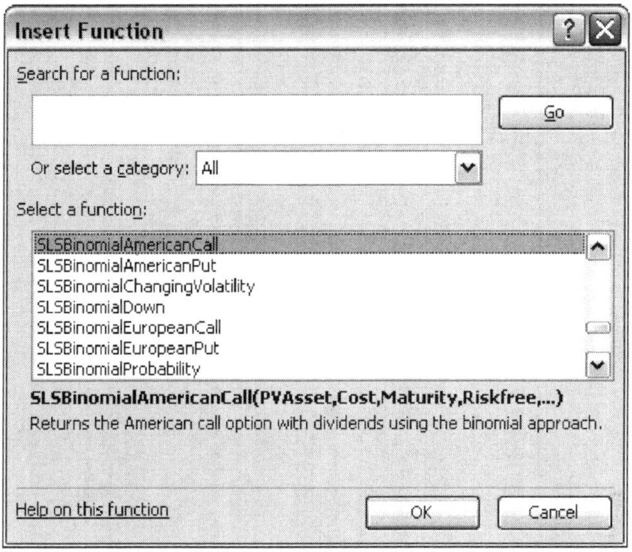

FIGURE 15.16 Excel's Equation Wizard.

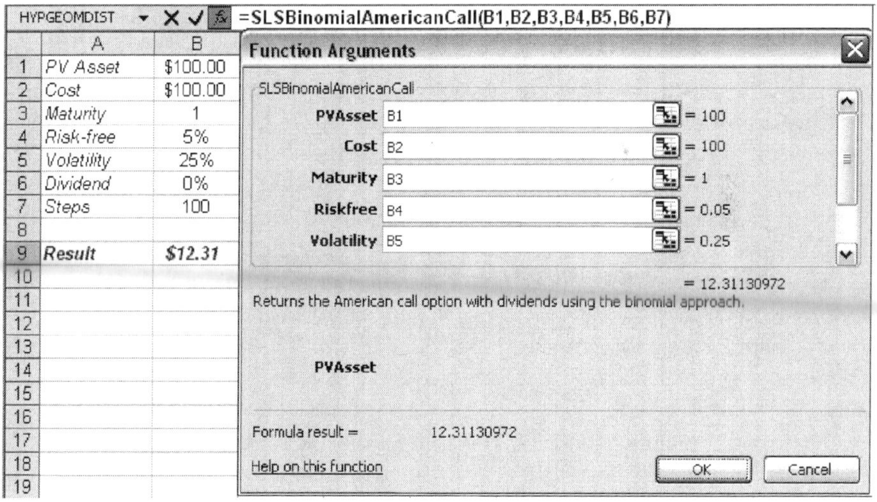

FIGURE 15.17 Using SLS Functions in Excel.

decision nodes indicating expected optimal times of execution of certain options is also available in this module. The results generated from this module are identical to those generated using the SLS and Excel functions, but have the added advantage of a visible lattice (lattices of up to 200 steps can be generated using this module). You are now equipped to start using the SLS software in building and solving real options, financial options, and employee stock options problems.

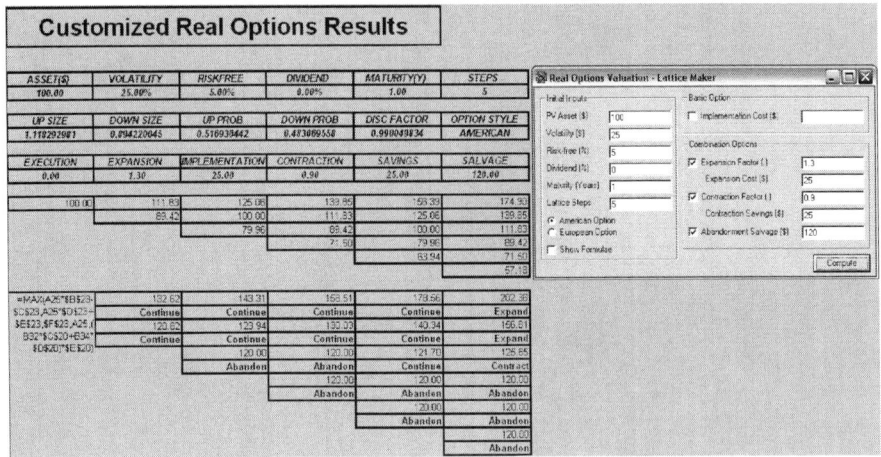

FIGURE 15.18 Lattice Maker.

15.7   ADVANCED EXOTIC OPTIONS VALUATOR

The Advanced Exotic Options Valuator is a comprehensive calculator of over 300 functions and models, from basic options to exotic options (e.g., from Black-Scholes to multinomial lattices to closed-form differential equations and analytical methods for valuing exotic options, as well as other options-related models such as bond options, volatility computations, delta–gamma hedging, and so forth). Figure 15.19 illustrates the valuator. You can click on the **Load Sample Values** button to load some samples to get started. Then, select the **Model Category** (left panel) as desired and select the **Model** (right panel) you wish to run. Click **Compute** to obtain the result. Please note that this valuator complements the ROV (Real Options Valuation, Inc.) Risk Modeler and ROV Valuator software tools, with over 800 functions and models, also developed by Real Options Valuation, Inc., which are capable of running at extremely fast speeds, handling large data sets, and linking into existing Open Database Connectivity (ODBC)-compliant databases (e.g., Oracle, SAP, Access, Excel, comma-separated values [CSV], and so forth). Finally, if you wish to access these 800 functions (including the ones in this Options Valuator tool), please use the ROV Modeling Toolkit software instead, where you can access these functions and more, and run Monte Carlo simulation on your models using ROV's Risk Simulator software.

15.7.1   Software Requirements, Key SLS Notes, and Tips

The Real Options SLS software imposes the following minimum requirements:

- Windows XP, Windows Vista, Windows 2007 or later
- Excel XP, Excel 2003, Excel 2007, 2010 or later
- .NET Framework 2.0 or later
- Administrative rights (during installation only)

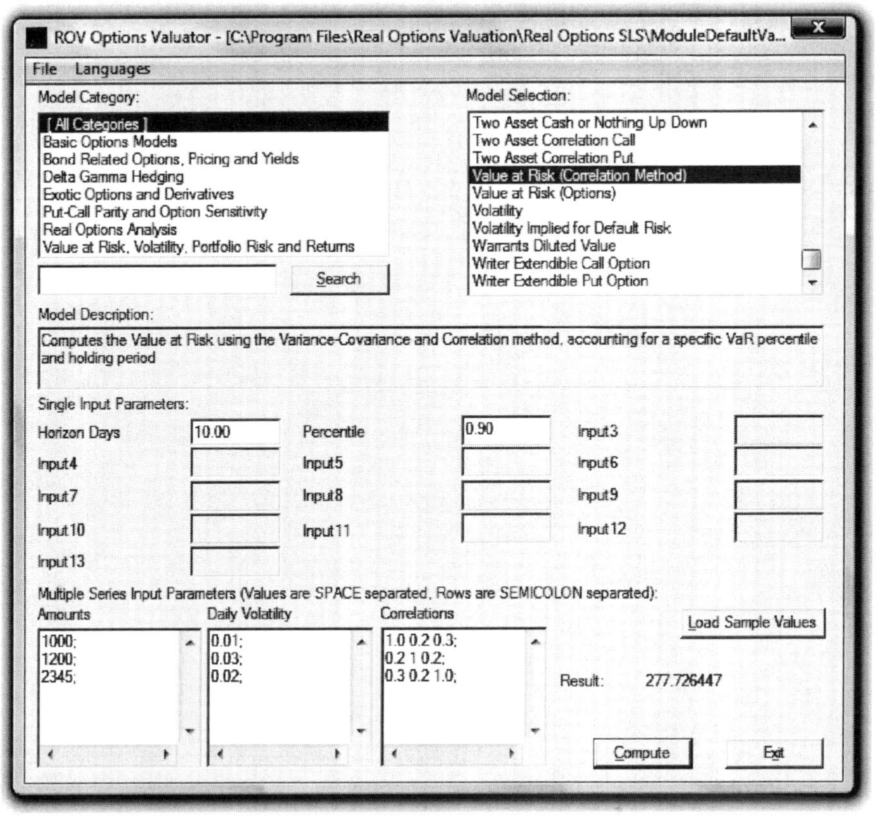

FIGURE 15.19 Exotic financial options valuator.

- 1GB of RAM or more
- 100 MB of free hard drive space

To install the software, make sure that your system has 1GB of RAM or more and 100 MB of free hard drive space. You need to first install .NET Framework 2.0 before proceeding with the Real Options SLS software installation. Next, install the Real Options SLS software by going to www.realoptionsvaluation.com, clicking on Downloads, and selecting Real Options SLS. The trial version is exactly the same as the full version except that it expires after ten days, during which you would need to obtain the full license or extension license to extend the use of the software. Install the software by following the onscreen prompts. If you have the trial version and wish to obtain the permanent license, visit **www.realoptionsvaluation.com** and click on the Purchase link (left panel of the Web site) and complete the purchase order. If you are purchasing or have already purchased the software, simply download and install the software.

There is a very detailed user's manual that comes with the software (simply start any of the modules and click on Help and select User Manual), where you will find additional detailed information about using the software, step-by-step instructions, a list of over 80 example models, and numerous case studies to get you started, complete with detailed instructions, sample numerical problems, and hands-on solutions.

REFERENCES

Mun, J. 2002. *Real Options Analysis: Tools and techniques for valuing strategic investments and decisions.* New York: John Wiley.
Mun, J. 2006. *Real Options Analysis: Tools and techniques,* 2nd ed. New York: Wiley Finance.

Index